MEMOIRS
of the
American Mathematical Society

ISSN 0065-9266 (print)
ISSN 1947-6221 (online)

Number 1469

Dehn Fillings of Knot Manifolds Containing Essential Twice-Punctured Tori

Steve Boyer
Cameron McA. Gordon
Xingru Zhang

March 2024 • Volume 295 • Number 1469 (first of 6 numbers)

AMERICAN MATHEMATICAL SOCIETY

Library of Congress Cataloging-in-Publication Data

Cataloging-in-Publication Data has been applied for by the AMS.
See http://www.loc.gov/publish/cip/.
DOI: https://doi.org/10.1090/memo/1469

Memoirs of the American Mathematical Society

This journal is devoted entirely to research in pure and applied mathematics.

Subscription information. Beginning in 2024, Memoirs will be published monthly through 2026. Memoirs is also accessible from www.ams.org/journals. The 2024 subscription begins with volume 293 and consists of twelve mailings, each containing one or more numbers. Individual subscription prices for 2024 are as follows. For electronic only: US$1013. For paper delivery: US$1,163. Add US$22 for delivery within the United States; US$150 for surface delivery outside the United States. Upon request, subscribers to paper delivery of this journal are also entitled to receive electronic delivery. For information on institutional pricing, please visit `https://www.ams.org/publications/journals/subscriberinfo`. Subscription renewals are subject to late fees. See `www.ams.org/journal-faq` for more journal subscription information. Each number may be ordered separately; please specify number when ordering an individual number.

Back number information. For back issues see `www.ams.org/backvols`.

Subscriptions and orders should be addressed to the American Mathematical Society, P. O. Box 845904, Boston, MA 02284-5904 USA. *All orders must be accompanied by payment.* Other correspondence should be addressed to 201 Charles Street, Providence, RI 02904-2213 USA.

Memoirs of the American Mathematical Society (ISSN 0065-9266 (print); 1947-6221 (online)) is published bimonthly (each volume consisting usually of more than one number) by the American Mathematical Society at 201 Charles Street, Providence, RI 02904-2213 USA. Periodicals postage paid at Providence, RI. Postmaster: Send address changes to Memoirs, American Mathematical Society, 201 Charles Street, Providence, RI 02904-2213 USA.

This publication is indexed in *Mathematical Reviews®*, *Zentralblatt MATH*, *Science Citation Index®*, *Science Citation Index™-Expanded*, *ISI Alerting Services℠*, *SciSearch®*, *Research Alert®*, *CompuMath Citation Index®*, *Current Contents®/Physical, Chemical & Earth Sciences*.
This publication is archived in *Portico* and *CLOCKSS*.
Printed in the United States of America.

∞ The paper used in this book is acid-free and falls within the guidelines established to ensure permanence and durability.
Visit the AMS home page at `https://www.ams.org/`

10 9 8 7 6 5 4 3 2 1 29 28 27 26 25 24

Contents

Abstract

We show that if a hyperbolic knot manifold M contains an essential twice-punctured torus F with boundary slope β and admits a filling with slope α producing a Seifert fibred space, then the distance between the slopes α and β is less than or equal to 5 unless M is the exterior of the figure eight knot. The result is sharp; the bound of 5 can be realized on infinitely many hyperbolic knot manifolds. We also determine distance bounds in the case that the fundamental group of the α-filling contains no non-abelian free group. The proofs are divided into the four cases F is a semi-fibre, F is a fibre, F is non-separating but not a fibre, and F is separating but not a semi-fibre, and we obtain refined bounds in each case.

Received by the editor April 12, 2020, and, in revised form, Febraury 22, 2021, and June 30, 2021.

Article electronically published on April 3, 2024.

DOI: https://doi.org/10.1090/memo/1469

2020 *Mathematics Subject Classification.* Primary 57M25, 57M50, 57M99.

The first named author was partially supported by NSERC grant RGPIN 9446-2008.

The second named author was partially supported by NSF grant DMS-0906276.

CHAPTER 1

Introduction

This is the third in a series of papers in which we investigate the following conjecture of the second named author ([**Gor98**, Conjecture 3.4]). Throughout we assume that M is a *hyperbolic knot manifold*. That is, M is a compact, connected, orientable 3-manifold with torus boundary whose interior has a complete finite volume hyperbolic structure.

CONJECTURE 1.1 ((C. McA. Gordon)). *Suppose that M is a hyperbolic knot manifold and α, β are slopes on ∂M such that $M(\beta)$ is a toroidal manifold and $M(\alpha)$ is a Seifert manifold. If $\Delta(\alpha, \beta) > 5$, then M is the figure eight knot exterior.*

Recall that if M is the exterior of the figure eight knot then $M(\gamma)$ is hyperbolic unless $\gamma \in \{\infty, 0, \pm 1, \pm 2, \pm 3, \pm 4\}$. Moreover, $M(\infty) \cong S^3$, $M(0)$ is a torus bundle, $M(\pm 4)$ are toroidal, and $M(\pm 1), M(\pm 2)$ and $M(\pm 3)$ are Seifert manifolds.

We note that 6 is the threshold distance characterising the figure eight knot exterior: there are infinitely many triples $(M; \alpha, \beta)$ where M is a hyperbolic knot manifold with Seifert and toroidal filling slopes α and β such that $\Delta(\alpha, \beta) = 5$. See Example 2.1.

Suppose that α and β are slopes on ∂M such that $M(\beta)$ is toroidal and $M(\alpha)$ is a Seifert manifold. In Proposition 3.3 of [**BGZ3**] we showed that $\Delta(\alpha, \beta) \leq 3$ if one of $M(\alpha)$ or $M(\beta)$ is reducible. Further, in Theorem 1.2 of [**BGZ3**] we showed that if $M(\alpha)$ is toroidal then $\Delta(\alpha, \beta) \leq 4$ (see Theorem 1.4 below). Thus we may assume that both $M(\alpha)$ and $M(\beta)$ are irreducible, and $M(\alpha)$ is atoroidal. In particular, $M(\alpha)$ is a *small Seifert manifold*, that is, it admits a Seifert structure with base orbifold of the form $S^2(a, b, c)$ where $a, b, c \geq 1$.

Since M is hyperbolic and $M(\beta)$ is toroidal, M contains an essential punctured torus F with non-empty boundary of slope β. Let m be the minimum value of $|\partial F|$ over all such F. In [**BGZ2**] we verified the conjecture in the case that $m \geq 3$ and M admits no punctured torus of boundary slope β which is a fibre or semi-fibre. In [**BGZ3**] we proved the conjecture when M contains an essential once-punctured torus. In the present paper we consider the case $m = 2$. Note that this is the case that arises in the ± 4-surgeries on the figure eight knot.

Here is our main result.

THEOREM 1.2. *Let M be a hyperbolic knot manifold which contains an essential twice-punctured torus F with boundary slope β, and let α be a slope on ∂M such that $M(\alpha)$ is an irreducible small Seifert manifold. If $\Delta(\alpha, \beta) > 5$ then M is the exterior of the figure eight knot. More precisely:*

(1) *If F is a fibre in M, then $\Delta(\alpha, \beta) \leq 3$.*
(2) *If F is a semi-fibre in M, then $\Delta(\alpha, \beta) \leq 4$.*
(3) *If F is non-separating in M, though not a fibre, then $\Delta(\alpha, \beta) \leq 5$.*

(4) *If F is separating in M, though not a semi-fibre, and if $\Delta(\alpha, \beta) > 5$, then M is the exterior of the figure eight knot.*

Theorem 1.2 combines with Proposition 3.3 and Theorems 1.2 and 1.3 of [**BGZ3**], and Theorem 2.7 of [**BGZ2**], to give:

COROLLARY 1.3. *Let M be a hyperbolic knot manifold that is not an m-punctured torus bundle, $m \geq 3$, or an m-punctured torus semi-bundle, $m \geq 4$. Then Conjecture 1.1 holds for M.*

The triples $(M; \alpha, \beta)$ where $M(\alpha)$ and $M(\beta)$ are toroidal and $\Delta(\alpha, \beta) \geq 4$ have been classified, by Gordon [**Go1**] when $\Delta(\alpha, \beta) \geq 6$ and by Gordon and Wu [**GW**] when $\Delta(\alpha, \beta) = 4$ or 5. Using this, the following result, which deals with the case where $M(\beta)$ is toroidal and $M(\alpha)$ is toroidal Seifert, was proved in [**BGZ3**]. Here N is the 3-chain link (shown in Figure 2.1) exterior (also called the *magic manifold*), and slopes are parametrised in the usual way for exteriors of links in S^3.

THEOREM 1.4 ([**BGZ3**, Theorem 1.2]). *Let M be a hyperbolic knot manifold with slopes α and β on ∂M such that $M(\beta)$ is toroidal and $M(\alpha)$ is a toroidal Seifert manifold. Then*

(1) *$\Delta(\alpha, \beta) \leq 4$, and*
(2) *$\Delta(\alpha, \beta) = 4$ if and only if $(M; \alpha, \beta) \cong (N(-1/2, -1/2); -4, 0)$.*

REMARK 1.5. There is a mistake in the proof of [**BGZ3**, Lemma 2.5], in the case of $N(-4) \cong M_3$. Namely, in the table just before Lemma 2.5, in the list of exceptional slopes for $N(-4)$, "$-1/2$" should be "$-5/2$". Similarly, in part (c) of Lemma 2.5, "$N(-4, -1/2, \gamma)$" should be "$N(-4, -5/2, \gamma)$", and "$\gamma = -1/2$" should be "no γ". Finally, in the first sentence after the proof of Lemma 2.5, "parts (a) and (c)" should be "part (a)", and the phrase "and $N(-4)$, respectively" should be deleted.

We also consider the case that $M(\alpha)$ is *very small*, in other words, its fundamental group does not contain a non-abelian free group. In this case, consideration of the JSJ decomposition of $M(\alpha)$ shows that it is either a torus bundle over the circle, a torus semi-bundle over an interval, or a Seifert manifold whose base orbifold has non-negative Euler characteristic. As in the proof of Theorem 1.4, the results of [**Go1**] and [**GW**] can be used to deal with the cases where $M(\alpha)$ is a torus bundle or semi-bundle. These are described in Theorems 1.6 and 1.7 below. In these theorems, Wh denotes the exterior of the Whitehead link (shown in Figure 2.1), with slopes again parametrised in the usual way for link exteriors in S^3.

THEOREM 1.6. *Let M be a hyperbolic knot manifold with slopes α and β on ∂M such that $M(\beta)$ is toroidal and $M(\alpha)$ is a torus bundle. Then*

(1) *$\Delta(\alpha, \beta) \leq 4$, and*
(2) *$\Delta(\alpha, \beta) = 4$ if and only if $(M; \alpha, \beta) \cong (Wh(n); 0, -4)$, $n \in \mathbb{Z}$, $n \neq -4, -3, -2, -1, 0$.*

In part (2) of Theorem 1.6, note that $Wh(1)$ is the figure eight knot exterior and $(Wh(1); 0, 4) \cong (Wh(1); 0, -4)$.

THEOREM 1.7. *Let M be a hyperbolic knot manifold with slopes α and β on ∂M such that $M(\beta)$ is toroidal and $M(\alpha)$ is a torus semi-bundle. Then*

(1) *$\Delta(\alpha, \beta) \leq 4$, and*

(2) $\Delta(\alpha, \beta) = 4$ *if and only if* $(M; \alpha, \beta) \cong (Wh(-4n/(2n+1); -4, 0)$, $n \in \mathbb{Z}$, $n \neq -1, 0$.

In Corollary 1.3 we had to exclude the cases where M is either an m-punctured torus bundle, $m \geq 3$, or an m-punctured torus semi-bundle, $m \geq 4$. If we assume that $M(\alpha)$ is very small Seifert, then we can include the latter case, and we can also include the former case if $M(\alpha)$ is of C- or D-type (whose definitions can be found at the beginning of §7.1).

THEOREM 1.8. *Let M be a hyperbolic knot manifold with slopes α and β on ∂M such that $M(\beta)$ is toroidal and $M(\alpha)$ is a very small Seifert manifold. If $M(\alpha)$ is not of C- or D-type, assume that M is not an m-punctured torus bundle, $m \geq 3$. Then $\Delta(\alpha, \beta) \leq 5$.*

Theorem 1.8 is best possible; see Example 2.2.

The proofs of the four parts of Theorem 1.2 are independent and are dealt with in separate parts of the paper.

Here is how the paper is organized.

In Chapter 2 we show that Theorem 1.2 is sharp by producing an infinite family hyperbolic knot manifolds each of which contains an essential twice punctured torus of boundary slope β and admits a small Seifert filling slope α such that $\Delta(\alpha, \beta) = 5$. Theorems 1.6 and 1.7 are proven in Chapter 3. Chapters 4 and 5 contain assumptions, reductions and background material to be applied later in the paper. In Chapter 6 we construct curves in the $PSL_2(\mathbb{C})$-character varieties of certain amalgams of triangle groups which will play an essential part in the proof of Theorem 1.2. The first two assertions of this theorem are proven in Chapters 7 and 8. The relative JSJ method for studying Dehn fillings introduced in [**BCSZ1**] and further developed in [**BGZ2**] is outlined in Chapter 9 and used to prove the third assertion of Theorem 1.2 in Chapter 10. Further background material for dealing with the last assertion is contained in Chapters 11 and 13 while Chapter 14 provides sufficient conditions for recognizing that M is the figure eight knot exterior when $\Delta(\alpha, \beta) > 5$. The proof of the final assertion of Theorem 1.2 is then dealt with in Chapters 12, 15, 17, 18, and 19. The proof of Theorem 1.8 is given in Chapter 20.

Acknowledgments

The authors would like to thank Bruno Martelli for graciously performing computer calculations related to an earlier version of the paper, and the referee for their careful reading of the manuscript and especially for pointing out flaws in our original proofs of Lemmas 7.5 and 8.2.

CHAPTER 2

Examples

In this chapter we examine the sharpness of the theorems in the introduction. We assume throughout that M is a hyperbolic knot manifold which contains an essential twice-punctured torus F of boundary slope β and α is a slope on ∂M such that $M(\alpha)$ either Seifert fibred or very small.

We use N to denote the exterior of the hyperbolic 3-chain link (shown in Figure 2.1) with the coordinates on ∂N as given in [**MP**].

FIGURE 2.1. The Whitehead link and the 3 chain link

The following example shows that 6 is the threshold distance characterising the figure eight knot exterior in that there are infinitely many hyperbolic knot manifolds which contain an essential twice punctured torus of boundary slope β and admit a small Seifert filling slope α such that $\Delta(\alpha, \beta) = 5$.

EXAMPLE 2.1. For each integer slope $n \neq 0, -1, -2, -3$, the manifold $N(-\frac{3}{2}, n)$ is a hyperbolic knot manifold. This fact follows from [**MP**, Theorems 1.1 and 1.2]. Now $N(-\frac{3}{2}, n, -\frac{5}{2})$ is a toroidal manifold and $N(-\frac{3}{2}, n, 0)$ is a small Seifert fibred manifold, by [**MP**, Table A.6]. So we just need to verify that $N(-\frac{3}{2}, n)$ contains an essential twice-punctured torus of boundary slope $-\frac{5}{2}$. To this end, first note that the manifold $N(-\frac{3}{2})$ is hyperbolic by [**MP**, Theorem 1.1]. Next note that $N(-\frac{3}{2}, -\frac{5}{2})$ and $N(-\frac{3}{2}, 0)$ are both toroidal manifolds ([**MP**, Table 1]). Hence by [**GW2**, Theorem 1.1], $N(-\frac{3}{2})$ is the exterior of the Whitehead sister link (shown in [**GW2**, Figure 7.1 (c)]) since it is the only hyperbolic link exterior of two boundary components which admits two toroidal Dehn fillings on one of its boundary components at distance 5. Furthermore by [**MP**, Table 1], $N(-\frac{3}{2}, -\frac{5}{2})$ has a unique essential torus T (up to isotopy) and this torus T splits the manifold into two pieces, one is Seifert fibred over a disk with two cone points and the other is Seifert fibred over an annulus with one cone point. Now by [**GW2**, Theorem 1.1], the essential torus T in $N(-\frac{3}{2}, -\frac{5}{2})$ can be arranged to intersect the third boundary torus of N in two components of slope $-\frac{5}{2}$. Moreover by the unique torus decomposition of the toroidal manifold $N(-\frac{3}{2}, n, -\frac{5}{2})$ given in [**MP**, Table A.6], one can see clearly that the torus T remains essential in $N(-\frac{3}{2}, n, -\frac{5}{2})$. As T intersects the boundary torus of $N(-\frac{3}{2}, n)$ in two components (minimally possible) of slope $-\frac{5}{2}$, its restriction on $N(-\frac{3}{2}, n)$ is an essential twice-punctured torus of boundary slope $-\frac{5}{2}$.

The following example shows that Theorem 1.8 is sharp.

EXAMPLE 2.2. Taking $n = -4$ or -5 in Example 2.1, the Seifert filling on $N(-\frac{3}{2}, n)$ with slope 0 has finite fundamental group (of T-type or I-type respectively) while the slope 0 has distance 5 to the essential twice-punctured torus of boundary slope $-\frac{5}{2}$. Note that $N(-\frac{3}{2}, -4)$ is the figure eight sister manifold.

CHAPTER 3

Proof of Theorems 1.6 and 1.7

By [**Go1**], there are four triples $(M; \alpha, \beta)$ with $M(\alpha)$ and $M(\beta)$ toroidal and $\Delta(\alpha, \beta) \geq 6$. In all cases one readily checks that neither $M(\alpha)$ nor $M(\beta)$ is a torus bundle or semi-bundle. For example, one may proceed as follows. The manifolds M are $Wh(1), Wh(-5), Wh(2)$ and $Wh(-5/2)$. Now $Wh \cong N(1)$, and with respect to the slope parametrizations described above, $Wh(r) \cong N(1, r+1)$. An examination of Tables A.2 and A.3 in [**MP**] shows that none of the toroidal fillings in question is a torus bundle or semi-bundle.

For $\Delta = 4$ or 5, the triples $(M; \alpha, \beta)$ with $M(\alpha)$ and $M(\beta)$ toroidal are determined in [**GW**]. There are 14 hyperbolic manifolds M_i, $1 \leq i \leq 14$, each with a pair of toroidal filling slopes α_i, β_i at distance 4 or 5, where M_1, M_2, M_3 and M_{14} have two (torus) boundary components, and the others, one. It is shown in [**GW**] that a hyperbolic knot manifold has two toroidal filling slopes α and β at distance 4 or 5 if and only if either $(M; \alpha, \beta) \cong (M_i; \alpha_i, \beta_i)$ for some $4 \leq i \leq 13$, or $(M; \alpha, \beta) \cong (M_i(\gamma); \alpha_i, \beta_i)$, for $i = 1, 2, 3$ or 14 and some slope γ on the second boundary component of M_i.

First, one sees from [**GW**, Lemma 22.2] that for $6 \leq i \leq 13$, neither $M_i(\alpha_i)$ nor $M_i(\beta_i)$ is a torus bundle or semi-bundle.

Next, the proof of [**BGZ3**, Lemma 2.3] shows that neither $M_4(\alpha_4)$ nor $M_4(\beta_4)$ is a torus bundle or semi-bundle.

For M_5, one can argue as follows. By [**L2**, Section 6], $M_5 \cong N(1, -1/3)$, the toroidal filling slopes being -4 and 1. Then [**MP**, Table A.4] shows that neither $N(1, -1/3, -4)$ nor $N(1, -1/3, 1)$ is a torus bundle or semi-bundle.

It remains to consider the cases M_1, M_2, M_3, and M_{14}. Recall that $M_1 \cong Wh \cong N(1), M_2 \cong N(-1/2)$, the exterior of the 2-bridge link with associated rational number 10/3, and $M_3 \cong N(-4)$, the exterior of the Whitehead sister link (see [**GW**] and [**MP**, Table A.1]).

For $M_1 \cong N(1)$, the toroidal slopes are -3 and 1 (with respect to the standard slope coordinates for N, used in [**MP**]). We must therefore determine the slopes γ such that $N(1, \gamma)$ is hyperbolic, i.e. $\gamma \notin \{\infty, -3, -2, -1, 0, 1\}$ (see [**MP**, Table A.1]), $N(-3, 1, \gamma)$ and $N(1, 1, \gamma)$ are toroidal, and one is a torus bundle or semi-bundle. An examination of Tables 2, 3 and 4 of [**MP**] shows that the only such slopes are (a) $\gamma = n \in \mathbb{Z}$, $n \neq -1, 1$, and (b) $\gamma = -(2n+1)/(2n+3)$, $n \in \mathbb{Z}$, $n \neq -2, -1$. Note that $N(-3, 1, \gamma)$ is a torus bundle in case (a) and a torus semi-bundle in case (b). Translating the slopes into the standard coordinates for Wh gives the examples in parts (2) of Theorems 1.6 and 1.7.

For $M_2 \cong N(-1/2)$ and $M_3 \cong N(-4)$ we do a similar analysis, the corresponding sets of exceptional slopes being $\{\infty, -4, -3, -2, -1, 0\}$ and $\{\infty, -3, -2, -1, -5/2, 0\}$, respectively, and toroidal slopes $\{-4, 0\}$ and $\{-5/2, 0\}$

(see [**MP**, Table A.1]). We conclude from Tables 2, 3 and 4 of [**MP**] that no triples $(M; \alpha, \beta)$ of the required type arise.

Finally we take care of M_{14}. It follows from the classification in [**GW**] of the hyperbolic knot manifolds with toroidal fillings at distance 4 that $M_{14} \cong Y$, where Y is the manifold defined by Lee in [**L1**]. The two toroidal fillings of Y are $Y(0) \cong Y(4) \cong Q(2,2) \cup Wh$, where $Q(2,2)$ is the Seifert manifold with base orbifold $D^2(2,2)$. If $Q(2,2) \cup Wh(\gamma)$ were a torus bundle or semi-bundle then γ would be an exceptional slope for Wh, and therefore would be in $\{\infty, -4, -3, -2, -1, 0\}$. But from [**MP**, Table A.1] (making the appropriate change of slope coordinates) we see that for no such γ is $Q(2,2) \cup Wh(\gamma)$ a torus bundle or semi-bundle.

CHAPTER 4

Initial assumptions and reductions

We assume throughout the paper that M is a hyperbolic knot manifold. A *slope* on the boundary of M is a ∂M-isotopy class of essential simple closed curves. Slopes correspond bijectively with $\pm$ pairs of primitive elements of $H_1(\partial M)$ in the obvious way and we shall often represent a slope by a primitive element of $H_1(\partial M)$. The *rational longitude* on ∂M is the unique slope λ_M having the property that it represents a torsion element of $H_1(M)$.

To each slope γ on ∂M we associate the γ-Dehn filling of M, denoted by $M(\gamma)$, obtained by attaching a solid torus V to M in such a way that the meridional slope of V is identified with γ. The resulting manifold is independent of all choices.

Given a closed, essential surface S in M, we use $\mathcal{C}(S)$ to denote the set of slopes δ on ∂M such that S compresses in $M(\gamma)$. A slope η on ∂M is called a *singular slope* for S if $\eta \in \mathcal{C}(S)$ and $\Delta(\gamma,\eta) \leq 1$ for each $\gamma \in \mathcal{C}(S)$. For instance, if $M(\eta)$ is Seifert fibred with hyperbolic base orbifold other than a 2-sphere with three cone points or a projective plane with two cone points, then η is a singular slope of some closed, essential surface in M ([**BGZ1**, Theorem 1.7]). A theorem of Wu ([**Wu1**]) shows that if S is a closed essential surface in M for which $\mathcal{C}(S) \neq \emptyset$, then there is a singular slope for S.

PROPOSITION 4.1 (cf. [**BGZ1**, Theorem 1.5]). *If η is a singular slope for some closed essential surface S in M, then for an arbitrary slope γ we have*

$$\Delta(\gamma,\eta) \leq \begin{cases} 1 & \text{if } M(\gamma) \text{ is either small Seifert or reducible} \\ 1 & \text{if } M(\gamma) \text{ is Seifert fibred and } S \text{ does not separate} \\ 2 & \text{if } M(\gamma) \text{ is toroidal and } \mathcal{C}(S) \text{ is infinite} \\ 3 & \text{if } M(\gamma) \text{ is toroidal and } \mathcal{C}(S) \text{ is finite.} \end{cases}$$

Consequently if $M(\gamma)$ is not hyperbolic, then $\Delta(\gamma,\eta) \leq 3$. $\diamondsuit$

PROPOSITION 4.2. *Suppose that F is an essential, properly embedded, twice-punctured torus of boundary slope β which caps-off to a compressible torus in $M(\beta)$. If γ is a slope on ∂M such that $M(\gamma)$ is not hyperbolic, then $\Delta(\gamma,\beta) \leq 3$. If $M(\gamma)$ is a small Seifert manifold, then $\Delta(\gamma,\beta) \leq 1$.*

PROOF. If F is non-separating the proposition is a special case of [**BGZ3**, Proposition 3.1]. Assume then that F is separating. The surface F splits M into two components which we denote by $X^{\pm}$, and ∂F splits ∂M into two annuli which we denote by $B^{\pm}$ so that $F \cup B^{\pm} = \partial X^{\pm}$. Let $\widehat{F}$ be the corresponding torus in $M(\beta)$ obtained by capping off ∂F with two disjoint meridian disks D_1 and D_2 of the filling solid torus V_β. The disks D_1 and D_2 cut V_β into two components $H^{\pm}$ with $H^{\pm}$ attached to $X^{\pm}$ along $B^{\pm}$ as a 2-handle. Let $\widehat{X}^{\pm} = X^{\pm} \cup_{B^{\pm}} H^{\pm}$. If ∂X^{ϵ}, $\epsilon \in \{+,-\}$, is compressible in X^{ϵ}, then by the handle addition lemma ([**Ja2**, Theorem 2]), $\widehat{F} = \partial \widehat{X}^{\epsilon}$ is incompressible in $\widehat{X}^{\epsilon}$. Hence by the assumption of the proposition, ∂X^ϵ is incompressible in X^ϵ for at least one ϵ, say $\epsilon = +$. Then ∂X^+ is also incompressible in M, since F is (cf. [**Ha**, §1.2(5), page 15]). Pushing ∂X^+ slightly into the interior of X^+, we obtain an embedded closed separating incompressible surface S in M. Since S contains an essential simple closed curve which is isotopic to the center circle of the annulus $B^+ \subset \partial M$, β is a singular slope by [**CGLS**, Theorem 2.4.3]. The proposition now follows from Proposition 4.1. $\diamondsuit$

Here is a list of assumptions that will hold throughout the paper.

ASSUMPTIONS 4.3.

(1) α is a slope on ∂M such that $M(\alpha)$ is Seifert fibred.
(2) $\beta \neq \alpha$ is a slope on ∂M of an essential, twice-punctured torus F.
(3) F caps-off to an essential torus in $M(\beta)$ (cf. Proposition 4.2).
(4) Neither α nor β is a singular slope of a closed essential surface in M (cf. Proposition 4.1).
(5) $M(\alpha)$ is an irreducible, atoroidal, small Seifert manifold with base orbifold $S^2(a, b, c)$ where $a, b, c \geq 1$ (cf. [**BGZ3**, Theorem 1.2 and Proposition 3.3(b)]).
(6) $M(\beta)$ is irreducible (cf. [**BGZ3**, Proposition 3.3(b)]).

Assumption 4.3(5) implies that $b_1(M)$, the first Betti number of M, is at most 2 and if it is 2, then $M(\alpha)$ fibres over the circle with fibre a horizontal surface.

Here are some additional notations which will be used throughout the paper. For a set Ω and a subset $\sigma \subset \Omega$, $\Omega - \sigma$ denotes the complement of σ in Ω. When Ω is a manifold and $\sigma \subset \Omega$ a submanifold, $\Omega \setminus \sigma$ denotes the manifold which is the metric completion of $\Omega - \sigma$ with respect to the path metric on $\Omega - \sigma$.

CHAPTER 5

Culler-Shalen theory

We collect various results on the $PSL_2(\mathbb{C})$-character variety of M which will be used in the paper. See [**CGLS**], [**LM**], [**BZ2**] and [**BCSZ2**] for the details.

Denote by $\mathcal{D}$ the abelian subgroup of $PSL_2(\mathbb{C})$ consisting of diagonal matrices and by $\mathcal{N}$ the subgroup consisting of those matrices which are either diagonal or have diagonal coefficients 0. Note that $\mathcal{D}$ has index 2 in $\mathcal{N}$ and any element in $\mathcal{N} - \mathcal{D}$ has order 2.

The action of $SL_2(\mathbb{C})$ on $\mathbb{C}^2$ descends to one of $PSL_2(\mathbb{C})$ on $\mathbb{C}P^1$. We call a representation ρ with values in $PSL_2(\mathbb{C})$ *irreducible* if the associated action on $\mathbb{C}P^1$ is fixed point free, otherwise we call it *reducible*. We call it *strictly irreducible* if the action has no invariant subset in $\mathbb{C}P^1$ with fewer than three points. Equivalently, ρ is strictly irreducible if it is irreducible and is not conjugate to a representation with values in $\mathcal{N}$.

Let Γ be a finitely generated group. The set $R_{PSL_2}(\Gamma)$ of representations of Γ with values in $PSL_2(\mathbb{C})$ admits the structure of a $\mathbb{C}$-affine algebraic set [**LM**] called the $PSL_2(\mathbb{C})$-*representation variety* of Γ. The action of $PSL_2(\mathbb{C})$ on $R_{PSL_2}(\Gamma)$ determines an algebro-geometric quotient $X_{PSL_2}(\Gamma)$ whose coordinate ring is $\mathbb{C}[R_{PSL_2}(\Gamma)]^{PSL_2(\mathbb{C})}$ and a regular map $t : R_{PSL_2}(\Gamma) \to X_{PSL_2}(\Gamma)$ [**LM**]. This quotient is called the $PSL_2(\mathbb{C})$-*character variety* of Γ. For $\rho \in R_{PSL_2}(\Gamma)$, we denote $t(\rho)$ by χ_ρ and refer to it as the *character* of ρ. If $\chi_{\rho_1} = \chi_{\rho_2}$ and ρ_1 is irreducible, then ρ_1 and ρ_2 are conjugate representations. We can therefore call a character χ_ρ reducible, irreducible, or strictly irreducible if ρ has that property.

When Γ is the fundamental group of a path-connected space Y, we write $R_{PSL_2}(Y)$ rather than $R_{PSL_2}(\pi_1(Y))$, $X_{PSL_2}(Y)$ rather than $X_{PSL_2}(\pi_1(Y))$, and refer to them respectively as the $PSL_2(\mathbb{C})$-representation variety of Y and $PSL_2(\mathbb{C})$-character variety of Y.

We call a curve in $X_{PSL_2}(\Gamma)$ *non-trivial*, respectively *strictly non-trivial*, if it contains an irreducible character, respectively strictly irreducible character. All but at most finitely many characters on a (strictly) non-trivial curve are (strictly) irreducible.

Each $\gamma \in \pi_1(M)$ determines an element f_γ of the coordinate ring $\mathbb{C}[X_{PSL_2}(M)]$ satisfying

$$f_\gamma(\chi_\rho) = (\text{trace}(\rho(\gamma)))^2 - 4$$

where $\rho \in R_{PSL_2}(M)$. Each $\delta \in H_1(\partial M) = \pi_1(\partial M)$ determines an element of $\pi_1(M)$ well-defined up to conjugation and therefore an element $f_\delta \in \mathbb{C}[X_{PSL_2}(M)]$. Similarly each slope δ on ∂M determines an element of $\pi_1(M)$ well-defined up to conjugation and taking inverse, and so defines $f_\delta \in \mathbb{C}[X_{PSL_2}(M)]$.

To each curve X_0 in $X_{PSL_2}(M)$ we associate a function

$$\| \cdot \|_{X_0} : H_1(\partial M; \mathbb{R}) \to [0, \infty)$$

characterized by the fact that for each $\delta \in H_1(\partial M)$ we have $\|\delta\|_{X_0} = \text{degree}(f_\delta : X_0 \to \mathbb{C})$. It was shown in [**CGLS**] that $\|\cdot\|_{X_0}$ is a seminorm, which we refer to as the *Culler-Shalen seminorm* of X_0.

Recall that X_0 admits an affine desingularisation $X_0^\nu \xrightarrow{\nu} X_0$ where ν is surjective and regular. Moreover, the smooth projective model $\widetilde{X}_0$ of X_0 is obtained by adding a finite number of ideal points to X_0^ν. Thus $\widetilde{X}_0 = X_0^\nu \cup \mathcal{I}(X_0)$ where $\mathcal{I}(X_0)$ is the set of ideal points of X_0. There are natural identifications between the function fields of X_0, X_0^ν, and $\widetilde{X}_0$. Thus to each $f \in \mathbb{C}(X_0)$ we have corresponding $f^\nu \in \mathbb{C}(X_0^\nu) = \mathbb{C}(X_0)$ and $\tilde{f} \in \mathbb{C}(\widetilde{X}_0) = \mathbb{C}(X_0)$ where $f^\nu = f \circ \nu = \tilde{f}|X_0^\nu$.

For $x \in \widetilde{X}_0$ and $\gamma \in \pi_1(M)$ we use $Z_x(\tilde{f}_\gamma)$, respectively $\Pi_x(\tilde{f}_\gamma)$, to denote the multiplicity of x as a zero, respectively pole, of $\tilde{f}_\gamma$. From the definition of $\|\cdot\|_{X_0}$ we see that for each $\delta \in H_1(\partial M)$ we have

$$\|\delta\|_{X_0} = \sum_{x \in \widetilde{X}_0} Z_x(\tilde{f}_\delta) = \sum_{x \in \mathcal{I}(X_0)} \Pi_x(\tilde{f}_\delta) \tag{5.1}$$

For each group G, an epimorphism $\varphi : \pi_1(M) \to G$ determines a closed injective morphism $\varphi^* : X_{PSL_2}(G) \to X_{PSL_2}(M)$, $\chi_\rho \mapsto \chi_{\rho\circ\varphi}$. In particular, we can identify $X_{PSL_2}(M(\beta))$ with an algebraic subset of $X_{PSL_2}(M)$.

Recall that the triple $(M, \alpha\ , \beta)$ always satisfies Assumptions 4.3. The purpose of this chapter is to prepare some results to be applied later to deal with a special case when $X_{PSL_2(\mathbb{C})}(M(\beta))$ is positive dimensional and contains a nontrivial curve X_0. Under this extra condition, $X_0 \subset X_{PSL_2}(M(\beta)) \subset X_{PSL_2}(M)$ yields a Culler-Shalen seminorm for which $\|\beta\|_{X_0} = 0$ (since f_β is constantly equal to 0 on X_0). In fact, if β^* is a dual slope to β (i.e. $\Delta(\beta, \beta^*) = 1$) and we set $s_{X_0} = \|\beta^*\|_{X_0}$, then for each slope δ on ∂M we have

$$\|\delta\|_{X_0} = \Delta(\delta, \beta)s_{X_0}. \tag{5.2}$$

It follows from the proof of [**Bo**, Proposition 6.2] that for each $x \in X_0^\nu$ and slope $\delta \neq \beta$ we have $Z_x(\widetilde{f}_\delta) \geq Z_x(\widetilde{f}_{\beta^*})$. Set

$$J_{X_0}(\alpha) = \{x \in \widetilde{X}_0 : Z_x(\widetilde{f}_\alpha) > Z_x(\widetilde{f}_\delta) \text{ for some } \delta \in H_1(\partial M) - \{0\}\}$$

Since X_0 is non-trivial, each element of X_0 is the character of a representation which is either irreducible, or has non-abelian image, or has image $\{\pm I\}$. (See [**Bo**, §2].) In the last case, we must have $b_1(M(\beta)) = 2$ (cf. [**Bo**, Proposition 2.8]), so $\beta = \lambda_M$, where λ_M is the rational longitude of M. (Note that as remarked right after Assumptions 4.3, $b_1(M)$ is at most 2.)

It is shown in [**BZ2**, Lemma 4.1] that the inverse image in $R_{PSL_2}(M)$ of a non-trivial irreducible curve $X_0 \subset X_{PSL_2}(M)$ has a unique 4-dimensional component R_0 which is conjugation invariant and maps onto X_0.

LEMMA 5.1. *Suppose that Assumptions* 4.3 *hold and let* $X_0 \subset X_{PSL_2}(M(\beta)) \subset X_{PSL_2}(M)$ *be a non-trivial irreducible curve. Fix* $x \in J_{X_0}(\alpha) \cap X_0^\nu$ *and fix* $\rho \in R_0$ *satisfying* $\nu(x) = \chi_\rho$.

(1) ρ *factors through* $\pi_1(M(\alpha)) \to \pi_1(S^2(a,b,c)) = \Delta(a,b,c)$ *and can be chosen to be either irreducible or have non-abelian image.*
(2) χ_ρ *is a simple point of* $X_{PSL_2}(M)$.

(3) *If $\nu(x)$ is irreducible and X_0 is strictly non-trivial, then*

$$Z_x(\tilde{f}_\alpha) = \begin{cases} 1 & \text{if the image of } \rho \text{ is a dihedral group of order } 6 \text{ or more} \\ 2 & \text{if } \rho \text{ is strictly irreducible} \end{cases}$$

PROOF. It follows from the discussion preceding the lemma, the condition $x \in J_{X_0}(\alpha) \cap X_0^\nu$ implies $Z_x(\tilde{f}_\alpha) > Z_x(\tilde{f}_{\beta^*})$. Thus it follows from part (1)(b) of [**Bo**, Proposition 6.2] that $\rho(\pi_1(\partial M))$ is a nontrivial finite cyclic group. In particular $\nu(x)$ is not the character of the trivial representation. It also follows from [**CGLS**, Proposition 1.5.4] that $\rho(\alpha) = \pm I$, and so ρ induces a homomorphism $\pi_1(M(\alpha)) \to PSL_2(\mathbb{C})$. As noted just before the statement of the lemma, we can choose ρ to be irreducible or have non-abelian image. In either case, it further factors through $\pi_1(S^2(a,b,c)) = \Delta(a,b,c)$ by Lemma 3.1 of [**BeBo**], thus proving (1). It then follows from [**BeBo**, Proposition and Theorem 1.1] that χ_ρ is a simple point of $X_{PSL_2}(M)$, so (2) holds. Finally, (3) is a consequence of the method of the proof of [**BZ2**, Lemma 6.1] combined with (2), which completes the proof. ◇

Set

$$\mathcal{N}_{X_0}(\alpha)$$
$$= \{x \in J_{X_0}(\alpha) \cap X_0^\nu \mid \nu(x) = \chi_\rho \text{ where } \rho \text{ is irreducible and takes values in } \mathcal{N}\}$$

PROPOSITION 5.2. *Suppose that Assumptions 4.3 hold and let X_0 be a curve in $X_{PSL_2}(M(\beta)) \subset X_{PSL_2}(M)$ and $\beta^* \in H_1(\partial M)$ be a dual class to β.*

(1) *If $\tilde{f}_{\beta^*}$ has poles at each ideal point of $\tilde{X}_0$ and $n > 1$ divides $\Delta(\alpha,\beta)$, then there is a point $x \in J_{X_0}(\alpha) \cap X_0^\nu$ such that $\rho(\pi_1(\partial M)) = \mathbb{Z}/n$ for any ρ with $\nu(x) = \chi_\rho$.*

(2) *If X_0 is strictly non-trivial and $s_{X_0} \neq 0$, then*

$$\Delta(\alpha,\beta) = 1 + \frac{1}{s_{X_0}}(2|J_{X_0}(\alpha)| - |\mathcal{N}_{X_0}(\alpha)|)$$

as long as $J_{X_0}(\alpha) \subset X_0^\nu$, $\nu(x)$ is irreducible for each $x \in J_{X_0}(\alpha)$ and no element of $\mathcal{N}_{X_0}(\alpha)$ corresponds to the character of a representation with image a dihedral group of order 4.

PROOF. Part (1) follows from part (2) of [**Bo**, Proposition 6.2].

Under the conditions of part (2) of this proposition, note that a point x of $\tilde{X}_0$ belongs to $J_{X_0}(\alpha)$ if and only if $Z_x(\tilde{f}_\alpha) > Z_x(\tilde{f}_{\beta^*})$. Thus

$$\begin{aligned}
&\Delta(\alpha,\beta)s_{X_0} \\
&= \|\alpha\|_{X_0} \\
&= \sum_{x\in\tilde{X}_0} Z_x(\tilde{f}_\alpha) \\
&= \sum_{x\in\tilde{X}_0} Z_x(\tilde{f}_{\beta^*}) + \sum_{x\in J_{X_0}(\alpha)} Z_x(\tilde{f}_\alpha) \\
&= s_{X_0} + \sum_{x\in J_{X_0}(\alpha)} Z_x(\tilde{f}_\alpha).
\end{aligned}$$

The formula in part (2) now follows from Lemma 5.1(3). ◇

CHAPTER 6

Bending characters of triangle group amalgams

6.1. Curves of characters of free products of cyclic groups

In this section we construct curves in the character varieties of certain amalgams of triangle groups to be used to prove the cases of Theorem 1.2 examined in Chapter 7, Section 10.2, Chapter 18, and Chapter 19.

Fix integers $q \geq p \geq 2$ and write $\mathbb{Z}/p * \mathbb{Z}/q = \langle a, b : a^p = b^q = 1 \rangle$. It was shown in Example 3.2 of [**BZ2**] that $X_{PSL_2}(\mathbb{Z}/p * \mathbb{Z}/q)$ is a disjoint union of a finite number of isolated points and $\lfloor \frac{p}{2} \rfloor \lfloor \frac{q}{2} \rfloor$ non-trivial curves, each isomorphic to a complex line. Explicit parametrisations of these curves can be given as follows. For integers j, k with $1 \leq j \leq \lfloor \frac{p}{2} \rfloor$ and $1 \leq k \leq \lfloor \frac{q}{2} \rfloor$, set

$$\lambda = e^{\pi i j/p}, \mu = e^{\pi i k/q}, \tau = \mu + \mu^{-1}.$$

For $z \in \mathbb{C}$ define $\rho_z \in R_{PSL_2}(\mathbb{Z}/p * \mathbb{Z}/q)$ by

$$\rho_z(a) = \pm \begin{pmatrix} \lambda & 0 \\ 0 & \lambda^{-1} \end{pmatrix}, \rho_z(b) = \pm \begin{pmatrix} z & 1 \\ z(\tau - z) - 1 & \tau - z \end{pmatrix}.$$

The characters of the representations ρ_z parameterize a curve

$$X(j,k) \subset X_{PSL_2}(\mathbb{Z}/p * \mathbb{Z}/q).$$

Moreover, it is shown in [**BZ2**, Example 3.2] that the holomorphic map

$$\Psi_{(j,k)} : \mathbb{C} \to X(j,k), z \mapsto \chi_{\rho_z}$$

is bijective if $j < \frac{p}{2}$ and $k < \frac{q}{2}$, and a 2-1 branched cover otherwise. Since $(\text{trace}(\rho_z(ab)))^2 - 4 = ((\lambda - \lambda^{-1})z + \lambda^{-1}\tau)^2 - 4$, the map

$$f_{ab} : X(j,k) \to \mathbb{C}, \chi_{\rho_z} \mapsto (\text{trace}(\rho_z(ab)))^2 - 4$$

has degree 1 if $j = \frac{p}{2}$ or $k = \frac{q}{2}$, and 2 otherwise.

LEMMA 6.1. *Fix a positive integer $d > 2$ and an element A of order d in $PSL_2(\mathbb{C})$.*

(1) *If $1 \leq j < \frac{p}{2}$ and $1 \leq k < \frac{q}{2}$, then there is an irreducible character $\chi_\rho \in X(j,k)$ such that $\rho(ab) = A$.*
(2) *If $j = \frac{p}{2}$ or $k = \frac{q}{2}$, then there is an irreducible character $\chi_\rho \in X(j,k)$ such that $\rho(ab) = A$ for all but at most one curve $X(j,k) \subset X_{PSL_2}(\mathbb{Z}/p * \mathbb{Z}/q)$.*
(3) (a) *If $(p,q,d) \neq (2,3,6)$, there is an irreducible representation $\rho : \mathbb{Z}/p * \mathbb{Z}/q \to PSL_2(\mathbb{C})$ such that $\rho(ab) = A$.*
    (b) *If $(p,q,d) \neq (2,3,6), (2,6,3), (2,4,4)$, there is an irreducible representation $\rho : \mathbb{Z}/p * \mathbb{Z}/q \to PSL_2(\mathbb{C})$ such that $\rho(a)$ has order p, $\rho(b)$ has order q, and $\rho(ab) = A$.*

PROOF. Fix a curve $X(j,k) \subset X_{PSL_2}(\mathbb{Z}/p * \mathbb{Z}/q)$ and consider the parametrisation $\Psi_{(j,k)} : \mathbb{C} \to X(j,k), z \mapsto \chi_{\rho_z}$ described above. If $\omega = (\text{trace}(A))^2 - 4$, then $\rho_z(ab)$ is conjugate to A if and only if $f_{ab}(\chi_{\rho_z}) = \omega$. Hence we must show that under the hypotheses of the lemma, we can find a $z \in \mathbb{C}$ such that ρ_z is irreducible and $f_{ab}(\chi_{\rho_z}) = \omega$. It is clear that ρ_z is reducible if and only if $z^2 - \tau z + 1 = 0$, or equivalently, $z = \mu, \mu^{-1}$.

Let g_ω be the polynomial $g_\omega(z) = f_{ab}(\chi_{\rho_z}) - \omega = ((\lambda - \lambda^{-1})z + \lambda^{-1}\tau)^2 - (\omega + 4)$. By construction, $\omega \in (-4, 0)$, and the reader will verify that this implies that g has simple roots. Thus if there is no irreducible character $\chi_\rho \in X(j,k)$ such that $\rho(ab) = A$, then the two roots of g_ω are μ and μ^{-1}. On the other hand, if z is a root of g_ω, then $\rho_z(ab)$ is conjugate to A in $PSL_2(\mathbb{C})$, so the eigenvalues of $\rho_z(ab)$ coincide with those of A, at least up to sign. Hence the eigenvalues of $\rho_\mu(ab)$ and $\rho_{\mu^{-1}}(ab)$ coincide up to sign. It follows that $\lambda\mu^{-1} \in \{\pm\lambda\mu, \pm\lambda^{-1}\mu^{-1}\}$. But this implies that either $\lambda^2 = -1$ or $\mu^2 = -1$, i.e. $j = p/2$ or $k = q/2$. Part (1) of the lemma thus holds.

Now assume that $j = \frac{p}{2}$, so $\lambda = i$. If there is no irreducible character $\chi_\rho \in X(\frac{p}{2}, k)$ such that $\rho(ab) = A$, then by the argument in the preceding paragraph the eigenvalues of A are $\pm i\mu_k, \pm i\mu_k^{-1}$. (We write μ_k here to underline the dependence of μ on k.) Suppose as well that there is no irreducible character $\chi_\rho \in X(\frac{p}{2}, k')$ such that $\rho(ab) = A$, for some $1 \leq k' \leq \lfloor \frac{q}{2} \rfloor$. Then the eigenvalues of A are $\pm i\mu_{k'}, \pm i\mu_{k'}^{-1}$. Thus $\mu_{k'} \in \{\pm\mu_k, \pm\mu_k^{-1}\}$. Hence $e^{2\pi i(k \pm k')/q} = 1$, or equivalently, $k' \equiv \pm k \pmod q$. Our constraints on k and k' then imply that $k' = k$.

The case that $k = \frac{q}{2}$ can be handled similarly, which completes the proof of (2).

It follows from (1) and (2) that if there is no irreducible representation $\rho : \mathbb{Z}/p * \mathbb{Z}/q \to PSL_2(\mathbb{C})$ such that $\rho(ab) = A$, then $p = 2$ and $\lfloor \frac{q}{2} \rfloor = 1$, so $q = 2$ or 3. The case $q = 2$ is easily ruled out using elementary properties of the dihedral group $\Delta(2, 2, d)$. Suppose then that $q = 3$. The proofs of (1) and (2) show that some eigenvalue of A is conjugate to $\pm\lambda\mu = \pm ie^{\pm\pi i/3}$. But then $d = 6$, so (3)(a) holds.

To prove part (3)(b), suppose that there is no irreducible representation $\rho : \mathbb{Z}/p * \mathbb{Z}/q \to PSL_2(\mathbb{C})$ such that $\rho(a)$ has order p, $\rho(b)$ has order q, and $\rho(ab) = A$. By (1) and (2), we see that $p = 2$. Further, (2) implies that the value of Euler's totient function at q is at most 2, so q is either $2, 3, 4$, or 6. We have already seen that the case that $q = 2$ can be ruled out and that $q = 3$ implies that $(p, q, d) = (2, 3, 6)$. If $q = 4$ or 6, the argument of the previous paragraph shows that $(p, q, d) = (2, 4, 4)$ or $(2, 6, 3)$ or $(2, 6, 6)$. Since $(2, 6, 6)$ is a hyperbolic triple, there is a discrete faithful irreducible representation $\rho : \Delta(2, 6, 6) \to PSL_2(\mathbb{C})$, and since there is a unique conjugacy class of elements of order 6 in $PSL_2(\mathbb{C})$, we can assume that $\rho(ab) = A$, a contradiction. This completes the proof. ◇

6.2. Amalgamated products

Fix positive integers $p_+ \leq q_+, p_- \leq q_-$ and d, each at least 2, and consider the triangle groups

$$\Delta(p_+, q_+, d) = \langle a_+, b_+ : a_+^{p_+} = b_+^{q_+} = (a_+b_+)^d = 1 \rangle$$

$$\Delta(p_-, q_-, d) = \langle a_-, b_- : a_-^{p_-} = b_-^{q_-} = (a_-b_-)^d = 1 \rangle$$

For each $\epsilon \in \{\pm\}$, $\mathbb{Z}/d \cong \langle a_\epsilon b_\epsilon \rangle \leq \Delta(p_\epsilon, q_\epsilon, d)$. Let $\psi : \langle a_+ b_+ \rangle \to \langle a_- b_- \rangle$ be an isomorphism. Then

$$\psi(a_+ b_+) = (a_- b_-)^s$$

where $1 \leq s < d$ and $\gcd(s, d) = 1$. We consider the amalgamated free product

$$\Delta(p_+, q_+, d) *_\psi \Delta(p_-, q_-, d)$$

LEMMA 6.2. *There is a homomorphism* $\rho : \Delta(p_+, q_+, d) *_\psi \Delta(p_-, q_-, d) \to PSL_2(\mathbb{C})$ *satisfying:*

(1) *The restriction* ρ_ϵ *of* ρ *to* $\Delta(p_\epsilon, q_\epsilon, d)$ *is irreducible for both values of* ϵ.
(2) *If* ρ_+ *is not faithful then either* (p_+, q_+, d) *is a Euclidean triple or* $(p_-, q_-, d) = (2, 3, 6)$. *Further,*
   (a) *if* (p_+, q_+, d) *is a Euclidean triple, then* ρ_+ *has image isomorphic to*
      - $\Delta(2, 3, 3)$ *when* (p_+, q_+, d) *is a permutation of* $(2, 3, 6)$ *or* $(3, 3, 3)$;
      - $\Delta(2, 2, 4)$ *when* (p_+, q_+, d) *is a permutation of* $(2, 4, 4)$;
   (b) *if* (p_+, q_+, d) *is a spherical or hyperbolic triple and* $(p_-, q_-, d) = (2, 3, 6)$, ρ_+ *can be taken to factor through a representation of* $\Delta(p_+, q_+, 3)$ *which is faithful if* $(p_+, q_+, 3)$ *is spherical or hyperbolic.*
(3) $\rho_+(\langle a_+ b_+ \rangle) \cong \begin{cases} \mathbb{Z}/d & \text{if } (p_\epsilon, q_\epsilon, d) \neq (2,3,6) \text{ for both } \epsilon \\ \mathbb{Z}/3 & \text{if } (p_\epsilon, q_\epsilon, d) = (2,3,6) \text{ for some } \epsilon \end{cases}$
(4) (a) *If* $d > 2$, *then* $\rho_-(a_-)$ *has order* p_- *and either* $\rho_-(b_-)$ *has order* q_- *or*
      - $(p_+, q_+, d) = (2, 3, 6), (p_-, q_-, d) = (2, 6, 6)$ *and* $\rho_-(b_-)$ *has order* 3,
      - $(p_+, q_+, d) = (3, 3, 3), (p_-, q_-, d) = (2, 6, 3)$ *and* $\rho_-(b_-)$ *has order* 3,
      - $(p_-, q_-, d) = (2, 4, 4)$ *and* $\rho_-(b_-)$ *has order* 2.
   (b) *If* $d = 2$, ρ_- *is a discrete faithful representation when* $(p_-, q_-, 2)$ *is a spherical or hyperbolic triple, or an epimorphism* $\Delta(3, 6, 2) \to \Delta(3, 3, 2)$, *or an epimorphism* $\Delta(4, 4, 2) \to \Delta(4, 2, 2)$.
(5) *If* ρ *is conjugate to a representation with values in* $\mathcal{N}$, *then* $(p_\epsilon, q_\epsilon, d)$ *is a permutation of either* $(2, 4, 4)$ *or some* $(2, 2, n)$ *for both values of* ϵ.

PROOF. We construct ρ by piecing together representations $\rho_+ : \Delta(p_+, q_+, d) \to PSL_2(\mathbb{C})$ and $\rho_- : \Delta(p_-, q_-, d) \to PSL_2(\mathbb{C})$ which agree on $\langle a_+ b_+ \rangle \equiv_\psi \langle a_- b_- \rangle$.

First suppose that (p_+, q_+, d) is spherical or hyperbolic. If

- $(p_-, q_-, d) \neq (2, 3, 6)$, we choose ρ_+ to be discrete and faithful;
- $(p_-, q_-, d) = (2, 3, 6)$ and $(p_+, q_+, 3)$ is spherical or hyperbolic, we take ρ_+ to be a discrete, faithful representation of $\Delta(p_+, q_+, 3)$;
- $(p_-, q_-, d) = (2, 3, 6)$ and $(p_+, q_+, 3)$ is Euclidean, then $(p_+, q_+, 3) = (2, 6, 3)$ or $(3, 3, 3)$. In either case we take ρ_+ to be an epimorphism $\Delta(p_+, q_+, 3) \to \Delta(2, 3, 3)$.

There are several cases to consider when (p_+, q_+, d) is a Euclidean triple:

- if $(p_+, q_+, d) = (3, 3, 3)$, we take ρ_+ to be an epimorphism $\Delta(3, 3, 3) \to \Delta(2, 3, 3) \subset SO(3) \subset PSL_2(\mathbb{C})$. Note that each element of order 3 in $\Delta(3, 3, 3)$ is sent to an element of order 3 in $\Delta(2, 3, 3)$.
- if (p_+, q_+, d) is a permutation of $(2, 4, 4)$, we take ρ_+ to be an epimorphism of $\Delta(p_+, q_+, d)$ onto $\Delta(2, 2, 4) \subset SO(3) \subset PSL_2(\mathbb{C})$ which send $a_+ b_+$ to an element of order d.

- if (p_+, q_+, d) is a permutation of $(2, 3, 6)$, we take ρ_+ to be an epimorphism of $\Delta(p_+, q_+, d)$ onto $\Delta(2, 3, 3)$

The reader will verify that assertion (3) holds with these choices.

Next we construct ρ_-.

In the case that $\rho_+(\langle a_+b_+\rangle) \cong \mathbb{Z}/d$ and $d > 2$, neither (p_+, q_+, d) nor (p_-, q_-, d) is $(2, 3, 6)$. Lemma 6.1(3) implies that if $(p_-, q_-, d) \neq (2, 6, 3), (2, 4, 4)$, we can find an irreducible representation $\rho_- : \Delta(p_-, q_-, d) \to PSL_2(\mathbb{C})$ such that $\rho_-(a_-)$ has order p_-, $\rho_-(b_-)$ has order q_-, and $\rho_-(a_-b_-) = \rho_+(a_+b_+)^m$ where m is an integer satisfying $sm + dl = 1$ for some integer l. So we have $\rho_-(a_-b_-)^s = \rho_+(a_+b_+)$. This identity shows that ρ_+ pieces together with ρ_- to form a homomorphism ρ satisfying assertions (1) through (4). If $(p_-, q_-, d) = (2, 6, 3)$ or $(2, 4, 4)$, we take ρ_- to be an epimorphism $\Delta(2, 6, 3) \to \Delta(2, 3, 3)$ or an epimorphism $\Delta(2, 4, 4) \to \Delta(2, 2, 4)$. A conjugate of ρ_- will then piece together with ρ_+ since there is a unique conjugacy class of elements of order 3 in $PSL_2(\mathbb{C})$ and a unique conjugacy class of elements of order 4.

In the case that $\rho_+(\langle a_+b_+\rangle) \cong \mathbb{Z}/2$, we choose ρ_- to be a discrete faithful representation when $(p_-, q_-, 2)$ is a spherical or hyperbolic triple. Otherwise we choose ρ_- to be an epimorphism $\Delta(3, 6, 2) \to \Delta(3, 3, 2)$ or an epimorphism $\Delta(4, 4, 2) \to \Delta(2, 4, 2)$. Since there is a unique conjugacy class of elements of order 2 in $PSL_2(\mathbb{C})$, a conjugate of ρ_- will piece together with ρ_+ to produce a homomorphism ρ satisfying assertions (1) through (4).

If $\rho_+(\langle a_+b_+\rangle) \ncong \mathbb{Z}/d$, then $\rho_+(\langle a_+b_+\rangle) \cong \mathbb{Z}/3$ and some $(p_\epsilon, q_\epsilon, d)$ is $(2, 3, 6)$. If

- $(p_+, q_+, d) = (2, 3, 6)$ and $(p_-, q_-, 3)$ is spherical or hyperbolic, we take ρ_- to be the composition of the obvious epimorphism $\Delta(p_-, q_-, 6) \to \Delta(p_-, q_-, 3)$ with a faithful representation $\Delta(p_-, q_-, 3) \to PSL_2(\mathbb{C})$;
- if $(p_+, q_+, d) = (2, 3, 6)$ and $(p_-, q_-, 3)$ is Euclidean, then $(p_-, q_-, 3) = (2, 6, 3)$ or $(3, 3, 3)$ and we take ρ_- to be the composition of the quotient homomorphism $\Delta(p_-, q_-, 6) \to \Delta(p_-, q_-, 3)$ with an epimorphism $\Delta(p_-, q_-, 3) \to \Delta(2, 3, 3)$;
- $(p_+, q_+, d) \neq (2, 3, 6)$, then $(p_-, q_-, d) = (2, 3, 6)$ and we take ρ_- to be an epimorphism $\Delta(2, 3, 6) \to \Delta(2, 3, 3)$.

As above, a conjugate of ρ_- pieces together with ρ_+ to yield a homomorphism satisfying assertions (1) through (4).

To complete the proof, we verify that assertion (5) holds.

If $(p_\epsilon, q_\epsilon, d)$ is a permutation of $(2, 3, 6)$ for some ϵ, then the image of ρ_ϵ is $\Delta(2, 3, 3)$, which does not conjugate into $\mathcal{N}$.

Suppose that $(p_\epsilon, q_\epsilon, d)$ is not a permutation of $(2, 3, 6)$ for either choice of ϵ. Then $\rho_+(\langle a_+b_+\rangle) \cong \mathbb{Z}/d$ and the reader will verify that the image of ρ_+ conjugates into $\mathcal{N}$ if and only if (p_+, q_+, d) is a permutation of $(2, 4, 4)$ or some $(2, 2, n)$.

Suppose that (p_-, q_-, d) is not a permutation of $(2, 4, 4)$ and $d > 2$. Then (3) and (4) show that $\rho_-(a_-)$ has order p_-, $\rho_-(b_-)$ has order q_-, and $\rho_-(a_-b_-)$ has order d. Then as $\Delta(p_-, q_-, d)$ is generated by any two of a_-, b_-, a_-b_-, the image of ρ_- is conjugate into $\mathcal{N}$ if and only if two p_-, q_-, d are 2. If $(p_-, q_-, 2)$ is not $(4, 4, 2)$ or $(2, n, 2)$, then our choice of ρ_- guarantees that its image cannot be conjugate into $\mathcal{N}$, which completes the proof. ◇

Fix a homomorphism $\rho : \Delta(p_+, q_+, d) *_\psi \Delta(p_-, q_-, d) \to PSL_2(\mathbb{C})$ and denote by ρ_ϵ the restriction of ρ to $\Delta(p_\epsilon, q_\epsilon, d)$. Let $Z(\rho(\langle a_+b_+\rangle))$ denote the centraliser of

$\rho(\langle a_+b_+\rangle)$ in $PSL_2(\mathbb{C})$ and $Z^0(\rho(\langle a_+b_+\rangle))$ its component of the identity. For each $S \in Z(\rho_+(\langle a_+b_+\rangle))$, define $\rho_S : \Delta(p_+,q_+,d) *_\psi \Delta(p_-,q_-,d) \to PSL_2(\mathbb{C})$ to be the homomorphism determined by the push-out diagram

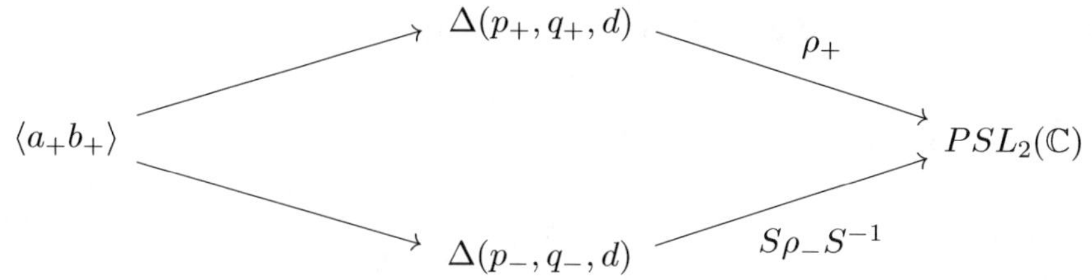

We say that the character χ_{ρ_S} is obtained by *bending* χ_ρ by S. The *bending function* of ρ is the map

$$\beta_\rho : Z^0_{PSL_2}(\rho(\langle a_+b_+\rangle)) \to X_{PSL_2}(\Delta(p_+,q_+,d) *_\psi \Delta(p_-,q_-,d)),\ S \mapsto \chi_{\rho_S}.$$

It is shown in [**BoiBo**, Lemma C.1] that β_ρ is constant if and only if one of the following two conditions holds:

- $\rho_+(\langle a_+b_+\rangle)=\{\pm I\}$ and either $\rho_+(\Delta(p_+,q_+,d))=\{\pm I\}$ or $\rho_-(\Delta(p_-,q_-,d)) = \{\pm I\}$;
- $\rho_+(\langle a_+b_+\rangle) \neq \{\pm I\}$ and either $\rho_+(\Delta(p_+,q_+,d)$ is abelian and reducible, or $\rho_-(\Delta(p_-,q_-,d))$ is abelian and reducible, or ρ is reducible.

In particular, if ρ is irreducible and β_ρ is constant, then $\rho_\epsilon(\Delta(p_\epsilon,q_\epsilon,d))$ is abelian and reducible for some ϵ.

We say that ρ can be *bent non-trivially* if β_ρ is non-constant. In this case, the image of β_ρ is contained in a curve in $X_{PSL_2}(\Delta(p_+,q_+,d) *_\psi \Delta(p_-,q_-,d))$.

LEMMA 6.3. *Let $\rho : \Delta(p_+,q_+,d) *_\psi \Delta(p_-,q_-,d) \to PSL_2(\mathbb{C})$ be an irreducible homomorphism as constructed in Lemma 6.2. Then*

(1) *ρ can be bent non-trivially to produce a non-trivial curve*
$$Y_0 \subset X_{PSL_2}(\Delta(p_+,q_+,d) *_\psi \Delta(p_-,q_-,d)).$$
(2) *Y_0 is strictly non-trivial if and only if $(p_\epsilon,q_\epsilon,d) \notin \{(2,2,d),(2,4,4)\}$ for some ϵ.*
(3) *Y_0 has exactly two ideal points unless $p_\epsilon = q_\epsilon = d = 2$ for some ϵ, in which case it has one. Further, for each integer k, the map $\tilde{f}_{c_+(a_+b_+)^kc_-}$ has a pole at each ideal point of Y_0 where $c_\pm \in \{a_\pm, a_\pm^{-1}, b_\pm, b_\pm^{-1}\}$.*

PROOF. By construction, $\rho_\epsilon = \rho|\Delta(p_\epsilon,q_\epsilon,d)$ is irreducible for both ϵ and so it can be bent non-trivially to produce a curve Y_0 in $X_{PSL_2}(\Delta(p_+,q_+,d) *_\psi \Delta(p_-,q_-,d))$ determined by bending ρ.

If Y_0 is *not* strictly non-trivial, it is a curve of characters of representations with values in $\mathcal{N}$. Then Lemma 6.2 implies that for each ϵ, $(p_\epsilon,q_\epsilon,d)$ is a permutation of either $(2,4,4)$ or a triple of the form $(2,2,n)$. Without loss of generality, we can assume that the images of both ρ_+ and ρ_- are contained in $\mathcal{N}$.

If, for some ϵ, we have $(p_\epsilon,q_\epsilon,d)=(2,n,2)$ for $n>2$ or $(4,4,2)$, then $\rho_+(a_+b_+) \in \mathcal{N} - \mathcal{D}$. After conjugating by an element of $\mathcal{D}$, we can assume that $\langle \rho_+(a_+b_+)\rangle = \{\pm I, \pm \left(\begin{smallmatrix} 0 & 1 \\ -1 & 0 \end{smallmatrix}\right)\}$. Then $Z^0(\rho(\langle a_+b_+\rangle)) = \{\pm \left(\begin{smallmatrix} x & y \\ -y & x \end{smallmatrix}\right) \mid x^2+y^2=1\}$. Let $\rho_{(x,y)}$ denote the representation obtained by piecing together ρ_+ and $\left(\begin{smallmatrix} x & y \\ -y & x \end{smallmatrix}\right)\rho_-\left(\begin{smallmatrix} x & y \\ -y & x \end{smallmatrix}\right)^{-1}$. Since

$(p_\epsilon, q_\epsilon) \neq (2,2)$, the only conjugate of $\rho_{(x,y)}$ which takes values in $\mathcal{N}$ is $\rho_{(x,y)}$ itself when $\epsilon = +$, and $\left(\begin{smallmatrix} x & y \\ -y & x \end{smallmatrix}\right)^{-1} \rho_{(x,y)} \left(\begin{smallmatrix} x & y \\ -y & x \end{smallmatrix}\right)$ otherwise. In either case, the fact that $\left(\begin{smallmatrix} x & y \\ -y & x \end{smallmatrix}\right)$ conjugates an irreducible subgroup of $\mathcal{N}$ into $\mathcal{N}$ implies that $\pm\left(\begin{smallmatrix} x & y \\ -y & x \end{smallmatrix}\right) \in \mathcal{N}$. Thus $x = 0$ or $y = 0$. It follows that for $xy \neq 0$, $\rho_{(x,y)}$ does not conjugate into $\mathcal{N}$, contrary to our assumptions. Thus $(p_\epsilon, q_\epsilon, d) \in \{(2,2,d), (2,4,4)\}$ for both ϵ. This proves the forward implication of (2).

For the reverse implication, suppose that (p_+, q_+, d) and (p_-, q_-, d) are of the form $(2,2,d)$ or $(2,4,4)$. As a first case, suppose that $(p_\epsilon, q_\epsilon, d)$ is either $(2,2,d)$ where $d > 2$ or $(2,4,4)$ for some ϵ. Lemma 6.2 then implies that the image of ρ_ϵ is isomorphic to $\Delta(2,2,d)$ for both ϵ, and we can suppose it is contained in $\mathcal{N}$. As $d > 2$, $\rho_+(a_+b_+) \in \mathcal{D}$. Hence $Z^0(\rho(\langle a_+b_+\rangle)) = \mathcal{D}$ and it is then easy to see that the image of ρ_S is contained in $\mathcal{N}$ for all $S \in \mathcal{D}$. Hence Y_0 contains no characters of strictly irreducible representations.

Suppose next that $(p_\epsilon, q_\epsilon, d) = (2,2,2)$ for both ϵ. Then we can conjugate ρ so that its image lies in $\mathcal{N}$ and $\rho_+(a_+b_+) \in \mathcal{D}$. Hence, again we have that Y_0 contains no characters of strictly irreducible representations, which completes the proof of (2).

Finally we consider assertion (3) for the case that $c_+ = a_+$ and $c_- = a_-$. The other cases are treated similarly.

Without loss of generality we suppose that $\rho_+(a_+b_+) \in \mathcal{D}$ and therefore $Z^0(\rho(\langle a_+b_+\rangle)) = \mathcal{D}$. There is a morphism

$$g : \mathbb{C}^* \to Y_0,\ t \mapsto \chi_{\rho_t}$$

where $\rho_{t-} = \left(\begin{smallmatrix} t & 0 \\ 0 & t^{-1} \end{smallmatrix}\right) \rho_- \left(\begin{smallmatrix} t^{-1} & 0 \\ 0 & t \end{smallmatrix}\right)$. Then

$$\rho_t(a_+(a_+b_+)^k a_-) = \rho_+(a_+)\rho_+(a_+b_+)^k \rho_{t-}(a_-).$$

By construction, ρ_+ and ρ_- are irreducible and $\rho_+(a_+b_+)$ is diagonal. Thus we can write

$$\rho_+(a_+) = \pm\begin{pmatrix} a & b \\ c & d \end{pmatrix} \quad \rho_+(a_+b_+) = \pm\begin{pmatrix} s & 0 \\ 0 & s^{-1} \end{pmatrix} \quad \rho_-(a_-) = \pm\begin{pmatrix} x & y \\ z & w \end{pmatrix}$$

where $bc \neq 0, yz \neq 0$ and $s \neq 0$. Then $\rho_+(a_+(a_+b_+)^k) = \pm\left(\begin{smallmatrix} a_0 & b_0 \\ c_0 & d_0 \end{smallmatrix}\right)$ where $b_0c_0 \neq 0$ and $\rho_t(a_-) = \pm\begin{pmatrix} x & t^2y \\ t^{-2}z & w \end{pmatrix}$ so that

$$\rho_t(a_+(a_+b_+)^k a_-) = \pm\begin{pmatrix} a_0x + t^{-2}b_0z & * \\ * & d_0w + t^2c_0y \end{pmatrix}$$

The trace of $\rho_t(a_+(a_+b_+)^k a_-)$ is therefore $\pm(c_0yt^2 + (a_0x + d_0w) + b_0zt^{-2})$, which diverges as t tends to either 0 or infinity since $b_0c_0yz \neq 0$. Thus the two ideal points of $\mathbb{C}^*$ are sent to ideal points of Y_0 and $\tilde{f}_{a_+(a_+b_+)^k a_-}$ has a pole at each of them. Thus Y_0 has at most two ideal points. It has only one if the two ideal points of $\mathbb{C}^*$ are sent to the same ideal point of Y_0.

To complete the proof, suppose that $g(t) = g(t')$. Then there is a $B \in PSL_2(\mathbb{C})$ such that $\rho_{t'} = B\rho_t B^{-1}$. In particular, $B\rho_+B^{-1} = \rho_+$ and $B\rho_-^t B^{-1} = \rho_-^{t'}$. Since

- ρ_+ is irreducible, either $B = \pm I$ or, up to conjugation, the image of ρ_+ is $\mathcal{K} = \{\pm I, \pm\left(\begin{smallmatrix} i & 0 \\ 0 & -i \end{smallmatrix}\right), \pm\left(\begin{smallmatrix} 0 & 1 \\ -1 & 0 \end{smallmatrix}\right), \pm\left(\begin{smallmatrix} 0 & i \\ i & 0 \end{smallmatrix}\right)\}$ and $B \in \mathcal{K}$.
- ρ_- is irreducible, either $B = \pm\begin{pmatrix} t'/t & 0 \\ 0 & t/t' \end{pmatrix}$ or, up to conjugation, the image of ρ_- is $\mathcal{K}$ and $\pm\begin{pmatrix} 1/t' & 0 \\ 0 & t' \end{pmatrix} B \left(\begin{smallmatrix} t & 0 \\ 0 & 1/t \end{smallmatrix}\right) \in \mathcal{K}$.

Hence if neither ρ_+ nor ρ_- has image isomorphic to $\Delta(2,2,2) \cong \mathcal{K}$, then $\pm I = B = \pm\begin{pmatrix} t'/t & 0 \\ 0 & t/t' \end{pmatrix}$. Hence $t' = \pm t$. It follows that the two ideal points of $\mathbb{C}^*$ are sent to distinct ideal points of Y_0.

On the other hand, suppose that $p_\epsilon = q_\epsilon = d = 2$ for some ϵ. By Lemma 6.2, the image of ρ_ϵ is isomorphic to $\Delta(2,2,2)$ and we can suppose that this image is $\mathcal{K}$. In this case, $\rho_+(a_+b_+) = \left(\begin{smallmatrix} i & 0 \\ 0 & -i \end{smallmatrix}\right) \in \mathcal{K}$. If $\epsilon = +$, the reader will verify that conjugation by $B = \pm\left(\begin{smallmatrix} 0 & 1 \\ -1 & 0 \end{smallmatrix}\right)$ takes ρ_t to $\rho_{\frac{1}{t}}$. Thus $g(t) = g(\frac{1}{t})$. It follows that g sends the two ideal points of $\mathbb{C}^*$ to the same ideal point of Y_0. Similarly, if $\epsilon = -$, conjugation by $B = \pm\begin{pmatrix} 0 & 1/t \\ -t & 0 \end{pmatrix}$ takes ρ_t to $\rho_{\frac{1}{t}}$. Thus $g(t) = g(\frac{1}{t})$, so again g sends the two ideal points of $\mathbb{C}^*$ to the same ideal point of Y_0. This completes the proof. ◇

6.3. HNN extensions

Fix integers $d, n \geq 2$ and consider the presentation $\Delta(d,n,d) = \langle a, b : a^d = b^n = (ab)^d = 1\rangle$. Let $\psi : \langle a \rangle \to \langle ab \rangle$ be an isomorphism. Then

$$\psi(a) = (ab)^s$$

where $1 \leq s \leq d-1$ and $\gcd(s,d) = 1$. Form the HNN extension

$$\Delta(d,n,d)*_\psi := \langle a, b, t : a^d = b^n = (ab)^d = 1, tat^{-1} = (ab)^s\rangle$$

A homomorphism $\rho : \Delta(d,n,d)*_\psi \to PSL_2(\mathbb{C})$ can be thought of as a pair $(\theta, \pm A)$ where $\theta = \rho|\Delta(d,n,d)$ and $\pm A = \rho(t) \in PSL_2(\mathbb{C})$ satisfies $A\theta(a)A^{-1} = \theta(ab)^s$. Conversely, a pair $(\theta, \pm A)$ where $\theta : \Delta(d,n,d) \to PSL_2(\mathbb{C})$ is a homomorphism and $A \in PSL_2(\mathbb{C})$ satisfies $A\theta(a)A^{-1} = \theta(ab)^s$ defines a homomorphism $\rho : \Delta(d,n,d)*_\psi \to PSL_2(\mathbb{C})$ in the obvious way. We shall write

$$\rho = (\theta, A)$$

Note that $(\theta, A), (\theta, B) \in R_{PSL_2}(\Delta(d,n,d)*_\psi)$ if and only if $B = AS$ for some $S \in Z_{PSL_2}(\rho(\langle a \rangle))$. Let $Z^0_{PSL_2}(\rho(\langle a \rangle))$ denote the component of the identity of $Z_{PSL_2}(\rho(\langle a \rangle))$. The map

$$\beta_{(\theta,A)} : Z^0_{PSL_2}(\rho(\langle a \rangle)) \to X_{PSL_2}(\Delta(d,n,d)*_\psi), S \mapsto \chi_{(\theta,AS)}$$

is called the *bending function* of (θ, A). It is shown in [**BoiBo**, Lemma C.3] that if $\beta_{(\theta,A)}$ is constant, then $\theta(\langle a \rangle) \neq \{\pm I\}$ and, after a possible conjugation, either

- $\theta(\Delta(d,n,d)) \subset \mathcal{D}$ and $A = \pm\left(\begin{smallmatrix} 0 & 1 \\ -1 & 0 \end{smallmatrix}\right)$, or
- $\theta(\Delta(d,n,d))$ and A are contained in the group of upper-triangular matrices.

In particular, if $\beta_{(\theta,A)}$ is constant, then θ is reducible and (θ, A) is either reducible or conjugate into $\mathcal{N}$.

We say that (θ, A) can be *bent non-trivially* if $\beta_{(\theta,A)}$ is non-constant. In this case, the image of $\beta_{(\theta,A)}$ is contained in a curve in $X_{PSL_2}(\Delta(d,n,d)*_\psi)$.

LEMMA 6.4. *There is a homomorphism $(\theta, A) : \Delta(d,n,d)*_\psi \to PSL_2(\mathbb{C})$ for which*

(1) *θ is irreducible.*
(2) *a, b and ab are sent to elements of order d, n and d respectively unless $(d,n) = (4,2)$. In this case a, b and ab are sent to elements of order 2.*

(3) (a) (θ, A) *can be bent non-trivially to produce a curve*

$$Y_0 \subset X_{PSL_2}(\Delta(d,n,d)*_\psi).$$

(b) Y_0 *is strictly non-trivial.*

(c) Y_0 *has two ideal points unless* $n = 2$ *and* $d \in \{2,4\}$ *(so* $\theta(\Delta(d,n,d)) \cong \Delta(2,2,2)$*), in which case it has one ideal point. For each* $j, l \in \mathbb{Z}$, *the function* $\tilde{f}_{ta^j ta^l}$ *has a pole at each ideal point of* Y_0.

PROOF. If (d,n,d) is a spherical triple, it is either $(2,n,2)$ or $(3,2,3)$. In either case we take θ to be injective with image $\Delta(d,n,d)$. Then $s \equiv \pm 1 \pmod{d}$, and as there is a unique conjugacy class of elements of order $d \in \{2,3\}$ in $PSL_2(\mathbb{C})$ we can find an $A \in PSL_2(\mathbb{C})$ such that $A\theta(a)A^{-1} = \theta(ab)^s$. Then (θ, A) satisfies (1), (2), and (3)(a) (cf. [**BoiBo**, Lemma C.3]).

If (d,n,d) is a Euclidean triple, it is either $(3,3,3)$ or $(4,2,4)$. In the first case we take θ to have image $\Delta(3,2,3)$ and note that each element of order 3 in $\Delta(3,3,3)$ is sent to an element of order 3 in $\Delta(3,2,3)$. In the second case we take θ to have image $\Delta(2,2,2)$. As in the previous paragraph, there is some $A \in PSL_2(\mathbb{C})$ for which (θ, A) is a homomorphism satisfying (1), (2), and (3)(a).

Assume then that (d,n,d) is a hyperbolic triple. In particular $d > 2$. Lemma 6.1 implies that we can find an irreducible representation $\theta : \mathbb{Z}/d * \mathbb{Z}/n \to PSL_2(\mathbb{C})$ such that $\theta(a)$ has order d, $\theta(b)$ has order n and $\theta(ab)$ is conjugate to $\theta(a)^m$, where m is a chosen integer such that $ms + dl = 1$ for some integer l. Then $\theta(a)$ is conjugate to $\theta(ab)^s$ and we can therefore construct a representation $(\theta, A) : \Delta(d,n,d)*_\psi \to PSL_2(\mathbb{C})$ satisfying (1), (2), and (3)(a).

Next we prove (3)(b). Suppose that the image of (θ, A) is conjugate into $\mathcal{N}$. Without loss of generality we can suppose that its image is contained in $\mathcal{N}$. Then $A \in \mathcal{N}$ and as θ is irreducible, part (2) of the lemma implies that either $d = 2$ or $(d,n,d) = (4,2,4)$. In the first case the image of θ is the dihedral group D_n while in the second it is D_2. In either case we can suppose that

$$\theta(\langle a \rangle) = \{\pm I, \pm \left(\begin{smallmatrix} 0 & 1 \\ -1 & 0 \end{smallmatrix}\right)\}$$

so, as we noted above, the component of the identity of the centraliser of $\theta(\langle a \rangle)$ is given by $\{\pm \left(\begin{smallmatrix} z & w \\ -w & z \end{smallmatrix}\right) : z, w \in \mathbb{C} \text{ and } z^2 + w^2 = 1\}$. Set $\rho_{(z,w)} = (\theta, A\left(\begin{smallmatrix} z & w \\ -w & z \end{smallmatrix}\right))$ and suppose that the image of $B\rho_{(z,w)}B^{-1}$ is contained in $\mathcal{N}$ for some $B \in PSL_2(\mathbb{C})$.

If $n > 2$, then $\rho_{(z,w)}(b) = \theta(b) \in \mathcal{D}$ and so as $B\rho_{(z,w)}(b)B^{-1} \in \mathcal{N}$, we must have $B\rho_{(z,w)}(b)B^{-1} = \theta(b)^{\pm 1}$. It follows that $B \in \mathcal{N}$. But then as $BA\left(\begin{smallmatrix} z & w \\ -w & z \end{smallmatrix}\right)B^{-1} \in \mathcal{N}$, so $\left(\begin{smallmatrix} z & w \\ -w & z \end{smallmatrix}\right) \in \mathcal{N}$ and therefore $z = 0$ or $w = 0$. Hence if $zw \neq 0$ and $n > 2$, $\rho_{(z,w)}$ does not conjugate to a representation with values in $\mathcal{N}$ and so Y_0 is strctly non-trivial.

Suppose that $n = 2$. Then the image of θ is D_2. If the image of $\rho_{(z,w)}$ contains a diagonal element of order different than 2, we can proceed as in the case $n > 2$ to complete the proof that Y_0 is strictly non-trivial. Otherwise the image of $\rho_{(z,w)}$ coincides with the image of θ, which we can take to be $\mathcal{K} = \{\pm I, \pm \left(\begin{smallmatrix} i & 0 \\ 0 & -i \end{smallmatrix}\right), \pm \left(\begin{smallmatrix} 0 & 1 \\ -1 & 0 \end{smallmatrix}\right), \pm \left(\begin{smallmatrix} 0 & i \\ i & 0 \end{smallmatrix}\right)\}$. Hence A is an element of the abelian group $\mathcal{K}$ and so $\theta(ab)^s = A\theta(a)A^{-1} = \theta(a)$. Since $\theta(a)$ has order 2 and s is odd, we have $\theta(ab) = \theta(a)$. But this is impossible as $\theta(a)$ and $\theta(ab)$ generate $\mathcal{K}$. This completes the proof of (3)(b).

Finally we prove (3)(c). We assume, without loss of generality, that $\theta(a)$ is diagonal, say

$$\theta(a) = \pm \begin{pmatrix} u & 0 \\ 0 & u^{-1} \end{pmatrix},$$

so $Z^0_{PSL_2}(\theta(\langle a \rangle)) = \mathcal{D} \cong \mathbb{C}^*$. Then there is a regular map

$$g : \mathbb{C}^* \to Y_0, r \mapsto \chi_{\rho_r}, \ \rho_r = (\theta, \pm A \left(\begin{smallmatrix} r & 0 \\ 0 & r^{-1} \end{smallmatrix}\right))$$

which factors through $\mathbb{C}^*/\{\pm 1\} \cong \mathbb{C}^*$.

Write $A = \pm \begin{pmatrix} x & y \\ z & w \end{pmatrix}$ and fix an integer k so that $sk \equiv 1 \pmod{d}$. Then as θ is irreducible and $A\theta(a)^k A^{-1} = \theta(ab)^{sk} = \theta(ab)$, the reader will verify that $x \neq 0$ and $w \neq 0$. A short calculation then shows that

$$\rho_r(ta^j ta^l) = \pm \begin{pmatrix} u^{(j+l)}x^2r^2 + u^{(l-j)}yz & * \\ * & u^{-(j+l)}w^2r^{-2} + u^{-(l-j)}yz \end{pmatrix}$$

Hence, up to sign, the trace of $\rho_r(ta^j ta^l)$ is $(u^{(j+l)}x^2)r^2 + (u^{(l-j)} + u^{-(l-j)})yz + (u^{-(j+l)}w^2)r^{-2}$, which diverges as r tends to either 0 or infinity since $x \neq 0$ and $w \neq 0$. Thus, each of the two ideal points of $\mathbb{C}^*$ is sent to an ideal point of Y_0 under the map g. Further, $\tilde{f}_{ta^j ta^l}$ has a pole at each of them.

If $g(r) = g(r')$, there is a $B \in PSL_2(\mathbb{C})$ such that $\rho_{r'} = B\rho_r B^{-1}$. In particular, $\theta = B\theta B^{-1}$ and $BA \left(\begin{smallmatrix} r' & 0 \\ 0 & (r')^{-1} \end{smallmatrix}\right) B^{-1} = A \left(\begin{smallmatrix} r & 0 \\ 0 & r^{-1} \end{smallmatrix}\right)$. Since θ is irreducible, either $B = \pm I$ or, up to conjugation, $\theta(\Delta(d,n,d)) = \mathcal{K}$ and $B \in \mathcal{K}$. In the former case, a simple calculation shows that $r' = \pm r$, while in the latter, (d,n) is either $(2,2)$ or $(4,2)$. In particular, if (d,n) is neither $(2,2)$ nor $(4,2)$, the two ideal points of $\mathbb{C}^*$ are sent to distinct ideal points of Y_0.

Suppose then that (d,n) is either $(2,2)$ or $(4,2)$ and therefore $\theta(\Delta(d,n,d)) = \mathcal{K}$. We have assumed that $\theta(a)$ is diagonal, so $\theta(a) = \pm \left(\begin{smallmatrix} i & 0 \\ 0 & -i \end{smallmatrix}\right)$, and up to conjugation by $\pm \left(\begin{smallmatrix} \sqrt{i} & 0 \\ 0 & 1/\sqrt{i} \end{smallmatrix}\right)$ we can suppose that $\theta(ab) = \pm \left(\begin{smallmatrix} 0 & 1 \\ -1 & 0 \end{smallmatrix}\right)$. It is easy to verify that if $v \neq 0$ and we set

$$A_v = \pm \left(\begin{smallmatrix} v & i/2v \\ iv & 1/2v \end{smallmatrix}\right) = \pm \left(\begin{smallmatrix} 1 & i/2 \\ i & 1/2 \end{smallmatrix}\right) \left(\begin{smallmatrix} v & 0 \\ 0 & 1/v \end{smallmatrix}\right),$$

then $A_v\theta(a)A_v^{-1} = \theta(ab) = \theta(ab)^s$, since s is odd. Further, $B = \pm \left(\begin{smallmatrix} 0 & i \\ i & 0 \end{smallmatrix}\right)$ conjugates (θ, A_v) to $(\theta, A_{\frac{1}{2v}})$. It follows that g sends the two ideal points of $\mathbb{C}^*$ to the same ideal point of Y_0. Thus Y_0 has a unique ideal point. $\diamondsuit$

CHAPTER 7

The proof of Theorem 1.2 when F is a semi-fibre

In this chapter we prove the semi-fibre case of Theorem 1.2.

PROPOSITION 7.1. *Suppose that M is a hyperbolic knot manifold which contains an essential twice-punctured torus F of boundary slope β and let α be a slope on ∂M such that $M(\alpha)$ is an irreducible small Seifert manifold. If F is a semi-fibre in M, then $\Delta(\alpha, \beta) \leq 4$.*

We divide the proof of Proposition 7.1 into two cases.

7.1. Proof of Proposition 7.1 when $M(\alpha)$ is very small

In this section we prove refined versions of the very small case of Proposition 7.1. In fact in this case, we allow the semi-fibre F to have arbitrary number of punctures.

Recall that the base orbifold of $M(\alpha)$ is $S^2(a, b, c)$ where a, b, c are positive integers. The condition that $M(\alpha)$ be very small corresponds to the requirement that either $\min\{a, b, c\} = 1$ or (a, b, c) is a Platonic or a Euclidean triple.

If $\min\{a, b, c\} = 1$ the fundamental group of $\pi_1(M(\alpha))$ is cyclic and we say that α is a *C-type filling slope.*

If (a, b, c) is a Platonic triple we say that α is, respectively, a *D-type, T-type, O-type* or *I-type filling slope* if up to permutation (a, b, c) is, respectively, $(2, 2, n), (2, 3, 3), (2, 3, 4)$, or $(2, 3, 5)$.

Finally, (a, b, c) is a Euclidean triple when (a, b, c) is one of $(2, 3, 6), (2, 4, 4)$, or $(3, 3, 3)$.

PROPOSITION 7.2. *Let M be a hyperbolic knot manifold that contains an m-punctured torus semi-fibre with boundary slope β, and let α be a slope on ∂M such that $M(\alpha)$ is an irreducible very small Seifert manifold. Then*

(1) *$M(\alpha)$ is not of T- or I-type.*
(2) *If $M(\alpha)$ is of C-type or has base orbifold $S^2(3, 3, 3)$ then $\Delta(\alpha, \beta) = 1$ and $m = 2$.*
(3) *If $M(\alpha)$ is of O-type or has base orbifold $S^2(2, 3, 6)$ then either*
    - *$\Delta(\alpha, \beta) = 1$ and $m = 2$ or 6, or*
    - *$\Delta(\alpha, \beta) = 3$ and $m = 2$.*
(4) *If $M(\alpha)$ has base orbifold $S^2(2, 4, 4)$ then either*
    - *$\Delta(\alpha, \beta) = 1$ and $m = 2, 4$ or 8, or*
    - *$\Delta(\alpha, \beta) = 2$ and $m = 2$ or 4, or*
    - *$\Delta(\alpha, \beta) = 4$ and $m = 2$.*
(5) *If $M(\alpha)$ is of D-type, then $\Delta(\alpha, \beta) \leq 3$.*

PROOF. The semi-fibre F separates M into two components X^+ and X^- each a twisted I-bundle over an $\frac{m}{2}$-punctured Klein bottle. Hence $\pi_1(F)$ is an index two subgroup of $\pi_1(X^\epsilon)$ for each $\epsilon \in \{\pm\}$ and there is an epimorphism $\varphi : \pi_1(M) \to \mathbb{Z}/2 * \mathbb{Z}/2$ whose kernel is $\pi_1(F)$. Clearly $\varphi(\beta) = 1$. We use x and y to denote the generators of the two $\mathbb{Z}/2$ factors. Let β^* be a dual slope to β on ∂M, then we may assume that each component of $\beta^* \cap X^\epsilon$ is an I-fibre of X^ϵ. Choose a disk region in F, which contains $F \cap \beta^*$, as a fat base point for each of $F, X^\epsilon, M, M(\beta), M(\alpha)$. Then each component of $\beta^* \cap X^\epsilon$ represents an element of $\pi_1(X^\epsilon)$ which is not contained in $\pi_1(F)$, and thus its image under φ is the generator of one of the $\mathbb{Z}/2$ factors. It follows that $\varphi(\beta^*) = (xy)^{m/2}$, at least up to exchanging x and y.

Write $\alpha = (\beta^*)^p \beta^q$ where $p = \Delta(\alpha, \beta)$. Then $\varphi(\alpha) = \varphi(\beta^*)^p = (xy)^{mp/2}$. It follows that φ induces a surjective homomorphism

$$\overline{\varphi} : \pi_1(M(\alpha)) \to (\mathbb{Z}/2 * \mathbb{Z}/2)/\langle\langle (xy)^{mp/2} \rangle\rangle = \langle x, y; x^2 = y^2 = (xy)^{mp/2} = 1\rangle = D_{mp/2}.$$

Set $n = mp/2$. As $H_1(D_n) \cong \mathbb{Z}/2$ (n odd) and $\mathbb{Z}/2 \oplus \mathbb{Z}/2$ (n even) while a very small Seifert manifold of type T or I has odd order first homology, part (1) of the proposition holds. When $n > 1$, D_n is an irreducible subgroup of $PSL_2(\mathbb{C})$. So if $M(\alpha)$ is of C-type, then $n = 1$ which means $p = \Delta(\alpha, \beta) = 1$ and $m = 2$. Hence when $\Delta(\alpha, \beta) > 1$, the base orbifold of $M(\alpha)$ is $S^2(a, b, c)$ with $a, b, c \geq 2$ and the surjective homomorphism $\overline{\varphi}$ must factor through $\pi_1(S^2(a, b, c)) = \Delta(a, b, c)$ ([**BeBo**, Lemma 3.1]). But then at least two of a, b, c are even as otherwise the abelianisation of $\Delta(a, b, c)$ is of odd order and so there is no surjective homomorphism $\Delta(a, b, c) \to D_n$. In particular $(a, b, c) \neq (3, 3, 3)$. So (2) holds.

In the case that α is of O-type (i.e. $(a, b, c) = (2, 3, 4)$) or $(a, b, c) = (2, 3, 6)$, then $n = 1$ or 3 since D_3 is the only irreducible dihedral group which is a quotient of either $\Delta(2, 3, 4)$ or $\Delta(2, 3, 6)$ (cf. [**BZ1**, Lemma 5.3], [**Bo**, Proposition 5.4]). This yields (3).

Similarly, if $(a, b, c) = (2, 4, 4)$, then $n = 1, 2$ or 4 (cf. [**Bo**, Proposition 5.3]), which gives (4).

Finally, suppose that α is of D-type. Then $(a, b, c) = (2, 2, l)$ for some $l \geq 2$ and therefore α is a Klein bottle filling slope on ∂M. By assumption β is also a Klein bottle filling slope. It then follows from [**MS**, Theorem 1.1] that $\Delta(\alpha, \beta) \leq 4$. Further, such triples $(M; \alpha, \beta)$ with $\Delta(\alpha, \beta) = 4$ are determined by Lee in [**L1**]. One checks that in all cases both $M(\alpha)$ and $M(\beta)$ are toroidal. Here are the details.

Lee ([**L1**]) defines two infinite families of hyperbolic knot manifolds:

- X_n, $n \in \mathbb{Z}$, each having a pair of Klein bottle Dehn filling slopes α_n and β_n with $\Delta(\alpha_n, \beta_n) = 4$,
- $Y(r)$, $r \notin \{\infty, 0, 4\}$, where $Y(r)$ is the r-Dehn filling of the exterior Y of a 2-component link in S^3 for which $Y(r)(0)$ and $Y(r)(4)$ contain Klein bottles,

and shows that any triple $(M; \alpha, \beta)$ such that $M(\alpha)$ and $M(\beta)$ contain Klein bottles with $\Delta(\alpha, \beta) = 4$ is homeomorphic to either $(X_n; \alpha_n, \beta_n)$ for some $n \in \mathbb{Z}$, or $(Y(r); 0, 4)$ for some $r \neq \infty, 0$, or 4.

Now X_n is obtained by Dehn filling along one boundary component of $M_2 \cong N(-1/2)$, the exterior of the 10/3 2-bridge link ([**L1**]). Comparing the parametrisations of the exceptional slopes $\{\infty, -2, -1, 0, 1, 2\}$ in [**L1**] and $\{\infty, -4, -3, -2, -1, 0\}$ in [**MP**, Table A.1], we see that $X_n \cong N(-1/2, (2n-5)/2) \cong N(-1/2, -(2n+3)/2)$, where the second homeomorphism follows from [**MP**, Proposition 1.5(4)]. An examination of Table A.8 and (for the case $N(-1/2, -5/2) \cong$

$N(-1/2, -3/2)$) Table A.4 in [**MP**] now shows that $\{\alpha_n, \beta_n\} = \{-4, 0\}$ and $X_n(\alpha_n)$ and $X_n(\beta_n)$ are toroidal.

Finally we consider the manifolds $Y(r)$. By [**L1**], $Y(0) \cong Y(4) \cong Q(2,2) \cup Wh$, and the Klein bottle fillings on $Y(r)$ are $Y(r)(0) \cong Y(0,r)$ and $Y(r)(4) \cong Y(0, 4-r)$ (see [**L1**, Lemma 2.2]). Now $Y(0,r) \cong Q(2,2) \cup Wh(r)$, with the induced parametrisation of slopes on Wh. From Lemma 2.4 of [**L1**] we have that $Wh(2) \cong Q(2,4)$ and $Wh(3) \cong Q(2,3)$, so from Table A.1 of [**MP**] it follows that this parametrisation of slopes on Wh is related to the standard parametrisation on $N(-1)$ by $Wh(r) \cong N(-1, r-3)$. Since $r \neq \infty$, Table A.1 of [**MP**] shows that $Wh(r)$ has incompressible boundary, and hence $Y(0,r)$ is toroidal. Part (5) is proved. $\diamondsuit$

We can refine the bounds given in Proposition 7.2 in certain situations, which we describe next.

The semi-fibre F splits M into two components X^+, X^- and if $\widehat{F}$ denotes the closed surface in $M(\beta)$ obtained by attaching disjoint meridian disks of the β-filling solid torus to F, then $\widehat{F}$ splits $M(\beta)$ into two twisted I-bundles $\widehat{X}^+$ and $\widehat{X}^-$ over the Klein bottle. Let ϕ_ϵ denote the slope on $\widehat{F}$ of the Seifert fibre structure on $\widehat{X}^\epsilon$ with base orbifold $D^2(2,2)$ and set

$$d = \Delta(\phi_+, \phi_-)$$

We consider constraints on d, $M(\alpha)$ when $\Delta(\alpha, \beta) > 1$. Given the latter condition, Proposition 7.2 implies that α has type D or O, or (a,b,c) is either $(2,3,6)$ or $(2,4,4)$.

PROPOSITION 7.3. *Suppose that F is an m-punctured semi-fibre and that α has type O, so $(a,b,c) = (2,3,4)$, or (a,b,c) is either $(2,3,6)$ or $(2,4,4)$. Suppose further that $d \neq 1$ and $\Delta(\alpha,\beta) > 1$. Then $(a,b,c) = (2,4,4)$ and $\Delta(\alpha,\beta) = 2$.*

PROOF. We saw in the proof of Proposition 7.2 that there is an epimorphism

$$\varphi_1 : \pi_1(M(\beta)) \to \mathbb{Z}/2 * \mathbb{Z}/2$$

whose kernel is $\pi_1(\widehat{F})$. Pulling back the curve $X(1,1) \subset X_{PSL_2}(\mathbb{Z}/2 * \mathbb{Z}/2)$ of Section 6.1 produces a curve

$$X_1 = (\varphi_1)^*(X(1,1)) \subset X_{PSL_2}(M(\beta)) \subset X_{PSL_2}(M)$$

More generally, for each positive integer d' dividing d we can construct a curve $X_{d'}$ in the character variety of M as follows.

Killing the fibre classes ϕ_+, ϕ_- in $\pi_1(M(\beta))$ yields a surjection

$$\varphi_d : \pi_1(M(\beta)) \to \Delta(2,2,d) *_{\mathbb{Z}/d} \Delta(2,2,d)$$

where in the case that $d = 0$ we take $\Delta(2,2,d)$ to be $\mathbb{Z}/2 * \mathbb{Z}/2$ and $\mathbb{Z}/d$ to be the central $\mathbb{Z}$ of index 2 in each of the $(\mathbb{Z}/2 * \mathbb{Z}/2)$-factors.

If $d \geq 2$ let $Y_d \subset X_{PSL_2}(\Delta(2,2,d) *_\psi \Delta(2,2,d))$ be a curve constructed by bending a representation $\Delta(2,2,d) *_\psi \Delta(2,2,d) \to PSL(2,\mathbb{C})$ which restricts to an irreducible representation with image D_d on each of the two copies of $\Delta(2,2,d)$. (See Lemma 6.3.) Set

$$X_d = (\varphi_d)^*(Y_d) \subset X_{PSL_2}(M(\beta)) \subset X_{PSL_2}(M)$$

Lemma 6.3(3) shows that X_d has exactly one ideal point if $d = 2$ and two if $d > 2$. Further, $\tilde{f}_{\beta^*}$ has a pole at each ideal point of X_d so that $s_{X_d} \geq 1$ if $d = 2$ and $s_{X_d} \geq 2$ if $d > 2$.

In the case that $d = 0$, one can use the method of proof of Lemmas 6.1, 6.2, and 6.3 to produce a non-trivial curve Y_0 in $X_{PSL_2}((\mathbb{Z}/2 * \mathbb{Z}/2) *_{\mathbb{Z}} (\mathbb{Z}/2 * \mathbb{Z}/2))$ obtained by bending an irreducible homomorphism $(\mathbb{Z}/2 * \mathbb{Z}/2) *_{\mathbb{Z}} (\mathbb{Z}/2 * \mathbb{Z}/2) \to PSL_2(\mathbb{C})$ for which $\rho(a_+ b_+)$ is an element of infinite order in $PSL_2(\mathbb{C})$. In particular, the image of each representation with character in Y_0 is irreducible and infinite. As above,

$$X_0 = (\varphi_0)^*(Y_0) \subset X_{PSL_2}(M(\beta)) \subset X_{PSL_2}(M)$$

has two ideal points and $\tilde{f}_{\beta^*}$ has a pole at each of them. Hence $s_{X_0} \geq 2$.

Similarly, if $d' > 1$ is a positive integer dividing d we can use the obvious quotient homomorphism

$$\pi_1(M(\beta)) \to \Delta(2,2,d) *_{\mathbb{Z}/d} \Delta(2,2,d) \to \Delta(2,2,d') *_{\mathbb{Z}/d'} \Delta(2,2,d')$$

to construct a curve

$$X_{d'} \subset X_{PSL_2}(M(\beta)) \subset X_{PSL_2}(M)$$

each of whose characters is irreducible. Further, $s_{X'_d} \geq 1$ if $d' = 2$ and $s_{X'_d} \geq 2$ if $d' > 2$.

By construction, the image of each representation whose character lies on some $X_{d'}$ contains a normal subgroup isomorphic to $\mathbb{Z}/d'$.

It is shown in [**BeBo**, Lemma 3.1] that if $\rho : \pi_1(M(\alpha)) \to PSL_2(\mathbb{C})$ has non-diagonalisable image and has a non-trivial character, then it factors through the triangle group $\Delta(a,b,c)$. Therefore by Lemma 5.3 of [**BZ1**] and Propositions 5.3 and 5.4 of [**Bo**], ρ has image

$$\text{image}(\rho) \cong \begin{cases} D_3 \text{ or } O_{24} = \Delta(2,3,4) & \text{if } \alpha \text{ is of } O\text{-type} \\ D_3 \text{ or } T_{12} = \Delta(2,3,3) & \text{if } (a,b,c) = (2,3,6) \\ D_2 \text{ or } D_4 & \text{if } (a,b,c) = (2,4,4) \end{cases}$$

In particular, the image of ρ is a finite group whose non-trivial elements have order $2, 3$, or 4.

No $X_{d'}$ contains the character of a representation with image T_{12} or O_{24}. This is clear when $d' = 1$ since neither of these groups is generated by two elements of order 2, and it is true when $d' > 1$ since neither T_{12} nor O_{24} contain non-trivial cyclic normal subgroups. Thus the only representations of $\pi_1(M(\alpha))$ whose characters lies on some $X_{d'}$ have image D_2, D_3, or D_4.

Each of the curves $X_{d'}$ we constructed above has $s_{X_{d'}} > 0$ and so as

$$\|\alpha\|_{X_{d'}} = \Delta(\alpha,\beta) s_{X_{d'}}$$

(cf. (5.2)) and $\Delta(\alpha,\beta) > 1$, $J_{X_{d'}}(\alpha) \neq \emptyset$ for each d'. Note as well that since $\tilde{f}_{\beta^*}$ has a pole at each ideal point of $X_{d'}$, so does $\tilde{f}_\alpha = \tilde{f}_{(\beta^*)^p}$ where $p = \Delta(\alpha,\beta) > 1$. Hence, any element of $J_{X_{d'}}(\alpha)$ is contained in $X^\nu_{d'}$ and if its image in $X_{d'}$ is the character χ_ρ, then χ_ρ is a smooth point of $X_{PSL_2}(M)$ (Lemma 5.1). In particular, $X_{d'}$ is the unique curve in $X_{PSL_2}(M)$ containing it. We know, again by Lemma 5.1, that $\rho(\alpha) = \pm I$, and so ρ induces irreducible homomorphism

$$\bar{\rho} : \pi_1(M(\alpha)) \to PSL(2,\mathbb{C})$$

Thus the image of ρ is D_2, D_3, or D_4. This excludes the possibility that $d' = 0$ as all representations with characters on X_0 have infinite image. It also excludes the possibility that $d' > 4$. Thus $d \in \{2,3,4\}$.

Since $d > 1$, there are at least two curves X_1 and X_d, and as $\Delta(\alpha, \beta) > 1$, each has $J_{X_{d'}}(\alpha) \neq \emptyset$ and therefore contains the character of a representation with dihedral image. This rules out the cases that (a,b,c) is either $(2,3,4)$ or $(2,3,6)$, since these groups admit a unique character of an irreducible representation with dihedral image. Thus $\Delta(\alpha, \beta) \leq 1$ when $d \neq 1$ and $(a,b,c) = (2,3,4)$ or $(2,3,6)$.

Finally suppose that $(a,b,c) = (2,4,4)$. Then $\pi_1(M(\alpha))$ admits exactly three irreducible characters of representations with values in $PSL(2,\mathbb{C})$, one with image D_2 and two with image D_4, by Proposition 5.3 of [**BeBo**]. Thus d is either 2 or 4, as is $\Delta(\alpha,\beta)$ by Proposition 5.2. Further, the latter result implies that if $\Delta(\alpha, \beta) = 4$, each of disjoint X_1, X_2 and X_4 contains at least two elements of $J_{X_{d'}}(\alpha)$, which is impossible. Thus $\Delta(\alpha, \beta) = 2$. $\diamondsuit$

7.2. Proof of Proposition 7.1 when $M(\alpha)$ is not very small

In this section we verify Proposition 7.1 in the case that $M(\alpha)$ is not very small.

There is a unique 2-fold cover $\widetilde{M} \to M$ where $\widetilde{M}$ is an F-bundle over S^1. Indeed, using the notation of the proof of Lemma 7.2, it corresponds to the composition of φ with the epimorphism $\mathbb{Z}/2 * \mathbb{Z}/2 \to \mathbb{Z}/2$ with kernel $\langle xy \rangle$. Let θ denote the covering involution of this cover. The reader will verify that $\partial\widetilde{M}$ has two boundary components T_1 and T_2, so the covering map restricts to a homeomorphism between T_i and ∂M for each i. Hence each slope γ on ∂M lifts to a slope γ_i on T_i $(i = 1,2)$ such that $\theta(\gamma_1) = \gamma_2$. Thus there is an induced 2-fold cover $\widetilde{M}(\gamma_1, \gamma_2) \to M(\gamma)$.

Pull back the Seifert structure on $M(\alpha)$ to $\widetilde{M}(\alpha_1, \alpha_2)$. The base orbifold $\mathcal{B}$ of $\widetilde{M}(\alpha_1, \alpha_2)$ covers that of $M(\alpha)$ with degree 1 or 2, and so is either $S^2(a,b,c)$ or $S^2(\frac{a}{2}, \frac{b}{2}, c, c)$, at least up to a permutation of a,b,c. Since $M(\alpha)$ is not very small, $\mathcal{B}$ has at least three cone points. In particular, $\widetilde{M}(\alpha_1, \alpha_2)$ is not a lens space.

LEMMA 7.4. *Suppose that M is a hyperbolic knot manifold which contains an essential twice-punctured torus F of boundary slope β and let α be a slope on ∂M such that $M(\alpha)$ is an irreducible small Seifert manifold. Suppose as well that F is a semi-fibre in M and $M(\alpha)$ is not very small. If $\Delta(\alpha,\beta) > 1$, then the cover $\mathcal{B} \to S^2(a,b,c)$ has degree 2 so for some permutation of a,b,c we can suppose that a and b are even and that $\mathcal{B} \cong S^2(a/2, b/2, c, c)$. Further, $c \geq 2\Delta(\alpha,\beta) - 2$.*

PROOF. Let $X_0 \subset X_{PSL_2}(M)$ be the $\mathbb{Z}/2 * \mathbb{Z}/2$-curve X_0 constructed in the proof of Lemma 7.2 . From the description of $\widetilde{M} \to M$ above, the image of X_0 in $X_{PSL_2}(\widetilde{M})$ is a curve of characters of representations which factor through the cyclic subgroup $\langle xy \rangle$ of $\mathbb{Z}/2 * \mathbb{Z}/2$. Hence the image is a curve of reducible characters. We saw in the proof of Lemma 7.2 that $J_{X_0}(\alpha) \subset X_0^\nu = X_0$ and that each element of $J_{X_0}(\alpha)$ is irreducible.

Suppose that $\Delta(\alpha,\beta) > 1$. If $\chi_\rho \in J_{X_0}(\alpha)$, then ρ factors through a homomorphism $\bar{\rho} : \pi_1(S^2(a,b,c)) \to PSL_2(\mathbb{C})$ (Lemma 5.1), and therefore the restriction of ρ to $\pi_1(\widetilde{M})$ factors through $\bar{\rho}|\pi_1(\mathcal{B})$. The image of ρ, and therefore that of $\bar{\rho}$, is a dihedral group D_n where $n \geq 2$ divides $\Delta(\alpha,\beta)$. Since ρ is irreducible, but $\rho|\pi_1(\widetilde{M})$ is reducible, $\rho(\pi_1(\widetilde{M})) = \bar{\rho}(\pi_1(\mathcal{B}))$ is the index 2 cyclic subgroup C_n of D_n.

Thus the cover $\mathcal{B} \to S^2(a,b,c)$ must have degree 2. It follows that after possibly permuting a, b, c, we can suppose that a and b are even and that $\mathcal{B} \to S^2(a,b,c)$ is the obvious cover $S^2(a/2, b/2, c, c) \to S^2(a,b,c)$. This proves the first assertion of the lemma.

In order to prove the second assertion, note that by the previous paragraph the generators of order a and b in $\pi_1(S^2(a,b,c))$ are sent to $D_n - C_n$. Thus $\bar{\rho}$ factors through the dihedral group $\Delta(2,2,c) \cong D_c$ in the obvious way. Hence, for each $\chi_{\rho'} \in J_{X_0}(\alpha)$, ρ' factors as a composition $\pi_1(M) \to \pi_1(M(\alpha)) \to \pi_1(S^2(a,b,c)) \to \Delta(2,2,c) \to PSL_2(\mathbb{C})$. Since there are $\lfloor \frac{c}{2} \rfloor$ characters of irreducible representations $\Delta(2,2,c) \to PSL_2(\mathbb{C})$ and the image of each one conjugates into $\mathcal{N}$, Proposition 5.1(3) implies that $\Delta(\alpha, \beta) \leq 1 + \lfloor \frac{c}{2} \rfloor$. Hence $c \geq 2\Delta(\alpha,\beta) - 2$, which completes the proof. ♢

We will make use of the involution τ_F depicted in Figure 7.1. It is central in the mapping class group of F. The quotient F/τ_F is the 2-orbifold $D^2(2,2,2,2)$.

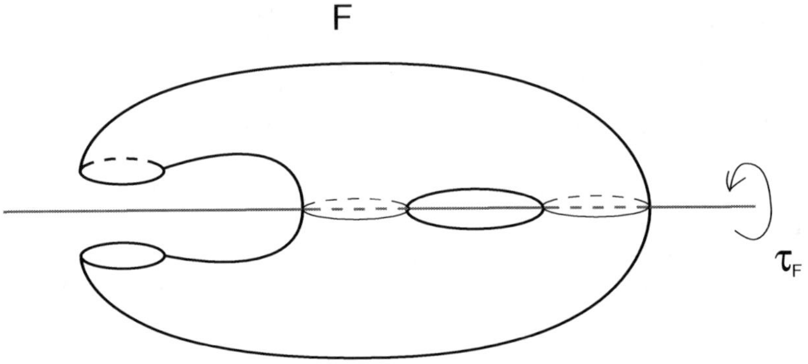

FIGURE 7.1. The involution τ_F

There is an involution τ on $\widetilde{M}$ induced by τ_F. The quotient $\widetilde{M}/\tau$ is homeomorphic to (V, L) where V is a solid torus and L is a 4-braid. We claim that for each slope γ on ∂M, τ induces an involution τ_γ of each $\widetilde{M}(\gamma_1, \gamma_2)$. To see this, note that τ is the lift of an involution τ_0 of M obtained by applying τ_F semi-fibre by semi-fibre. Then τ_0 restricts to a hyperelliptic involution of ∂M, and so acts on $H_1(\partial M)$ by multiplication by -1. Since τ is a lift of τ_0, the isomorphism $\tau_* : H_1(T_1) \to H_1(T_2)$ is multiplication by -1 under the identification induced by θ. In particular, τ sends the slope γ_1 to γ_2, which is what we needed to prove.

The restriction of the quotient map $\widetilde{M} \to V$ to each T_i yields a homeomorphism $T_i \to \partial V$ such that for any slope γ on ∂M, γ_1 and γ_2 determine the same slope $\bar{\gamma}$ on ∂V. Clearly $\widetilde{M}(\gamma_1, \gamma_2)/\tau_\gamma = V(\bar{\gamma})$. Further, the branch set $L_\gamma \subset V(\bar{\gamma})$ of this quotient is $L \subset V \subset V(\bar{\gamma})$.

Note that $\bar{\beta}$ is a meridian of V while if β^* is a dual class to β, $\bar{\beta}^*$ is a longitude of V. Thus after possibly changing the sign of α we can write

$$\alpha = p\beta^* + q\beta \tag{7.1}$$

where p, q are coprime. After possibly changing the signs of β^* and β we may assume that

$$p = \Delta(\alpha, \beta) \tag{7.2}$$

By construction

$$V(\bar{\alpha}) \cong L(p,q) \tag{7.3}$$

As in the proof of [**BGZ3**, Lemma 4.1], we can assume that τ_α preserves the Seifert structure on $\widetilde{M}(\alpha_1, \alpha_2)$.

LEMMA 7.5. *Suppose that M is a hyperbolic knot manifold which contains an essential twice-punctured torus F of boundary slope β and let α be a slope on ∂M such that $M(\alpha)$ is an irreducible small Seifert manifold. Suppose as well that F is a semi-fibre in M and $M(\alpha)$ is not very small. If τ_α reverses the orientation of the Seifert fibres of $\widetilde{M}(\alpha_1, \alpha_2)$, then $\Delta(\alpha, \beta) \leq 4$.*

PROOF. Suppose that τ_α reverses the orientation of the fibres of $\widetilde{M}(\alpha_1, \alpha_2)$. The method of proof of [**BGZ3**, Lemma 4.4] shows that either

(a) $\widetilde{M}(\alpha_1, \alpha_2)$ has base orbifold $S^2(p,p,m)$ for some $m \geq 2$ where the branch set L_α in $L(p,q)$
is a closed m/n-rational tangle in a Heegaard solid torus as shown in Figure 7.2 (the example
shown in Figure 7.3 should suffice to illustrate our convention for rational tangles), or

(b) $\widetilde{M}(\alpha_1, \alpha_2)$ has base orbifold $S^2(p,p,m,s)$ for some $m, s \geq 2$ where the branch set L_α in
$L(p,q)$ is a closed two tangle Montesinos link with rational tangles m/n and s/t.

Pull L_α back to the universal cover S^3 of $L(p,q)$ and call the resulting link $\widetilde{L}_\alpha$. Then $\widetilde{L}_\alpha$ is a Montesinos link with $r \geq p$ rational tangles. The 2-fold cover $\Sigma_2(\widetilde{L}_\alpha)$ of S^3 branched over $\widetilde{L}_\alpha$ is Seifert with base orbifold a 2-sphere with r cone points. By [**S**, Theorem 5.3] its Heegaard genus is realized by a vertical splitting, of genus $r-1$, unless it is as in case (2) of that theorem, in which case r is even and ≥ 4 and its Heegaard genus is realized by a horizontal splitting of genus $r-2$. On the other hand, L_α is the image of L in $V(\bar{\alpha})$, so $\widetilde{L}_\alpha$ is the closure of a 4-braid. Hence the Heegaard genus of $\Sigma_2(\widetilde{L}_\alpha)$ is at most 3. Hence either $r-1 \leq 3$, or $r \geq 4$ is even and $r-2 \leq 3$. It follows that $\Delta(\alpha, \beta) = p \leq r \leq 4$. ◇

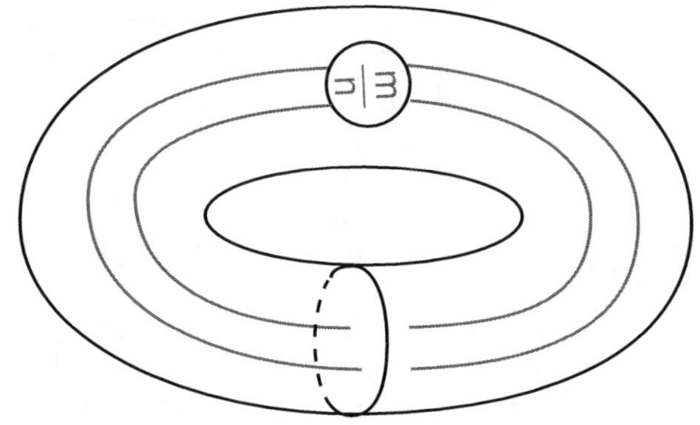

FIGURE 7.2. The branch set

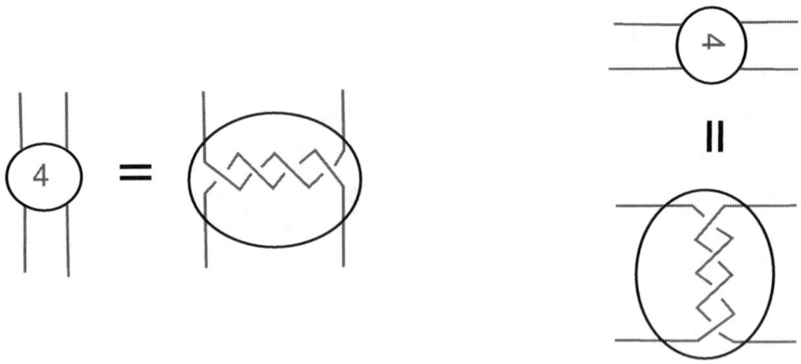

FIGURE 7.3. The convention for rational tangles

For the rest of this section we suppose that τ_α preserves the orientation of the fibres of $\widetilde{M}(\alpha_1, \alpha_2)$. As in [**BGZ3**, Lemma 4.3] we have

LEMMA 7.6. *Suppose that M is a hyperbolic knot manifold which contains an essential twice-punctured torus F of boundary slope β and let α be a slope on ∂M such that $M(\alpha)$ is an irreducible small Seifert manifold. Suppose as well that F is a semi-fibre in M and $M(\alpha)$ is not very small. If τ_α preserves the orientation of the Seifert fibres of $\widetilde{M}(\alpha_1, \alpha_2)$ and $\Delta(\alpha, \beta) \geq 4$, then*

1. *there is an induced Seifert structure on $L(p,q) \cong \widetilde{M}(\alpha_1, \alpha_2)/\tau_\alpha$ whose branch set L is a union of at most four Seifert fibres.*
2. *the induced Seifert structure on $L(p,q)$ has at least one cone point of order $\Delta(\alpha, \beta) - 1$.*
3. *L contains at least one regular fibre of $L(p,q)$ and if only one, $L(p,q)$ has no exceptional fibre of multiplicity 2.*

PROOF. By Lemma 7.4, $\widetilde{M}(\alpha_1, \alpha_2)$ has base orbifold $S^2(\bar{a}, \bar{b}, c, c)$ and $c \geq 2(\Delta(\alpha, \beta) - 1)$. As $M(\alpha)$ is not very small, $S^2(\bar{a}, \bar{b}, c, c)$ is hyperbolic.

Let $\bar{\tau}_\alpha$ be the map of $S^2(\bar{a}, \bar{b}, c, c)$ induced by τ_α. Either $\bar{\tau}_\alpha$ is the identity or it is an orientation-preserving involution with exactly two fixed points. In the former case, a fibre of multiplicity j in $\widetilde{M}(\alpha_1, \alpha_2)$ is sent to a fibre of multiplicity $\bar{j} = \frac{j}{\gcd(j,2)}$ in $L(p,q)$. Further, the fixed point set of τ_α is the union of exceptional fibres of $\widetilde{M}(\alpha_1, \alpha_2)$ of even multiplicity. Hence L has at most four components. Further, the base orbifold $\overline{\mathcal{B}}$ of $L(p,q)$ in the induced Seifert structure is $S^2(\bar{\bar{a}}, \bar{\bar{b}}, \bar{c}, \bar{c})$. Since $\bar{c} \geq \Delta(\alpha, \beta) - 1 \geq 3$ (Lemma 7.4) we must have $\bar{\bar{a}} = \bar{\bar{b}} = 1$. Thus $a, b \in \{2, 4\}$ and $\overline{\mathcal{B}} = S^2(\bar{c}, \bar{c})$. Further, since $S^2(a, b, c)$ is hyperbolic, at least one of a, b is 4. It follows that L contains at least one regular fibre. If there is only one, then $a = 2$ (say) and $b = 4$. It is easy to see that the lemma holds in this situation.

Next suppose that $\bar{\tau}_\alpha$ is not the identity. In this case L has at most two components. If $a = 2$, then $\bar{b} > 1$ and $\mathcal{B} = S^2(\bar{b}, c, c)$. As $\bar{\tau}_\alpha$ is not the identity, it fixes exactly one of the cone points of $\mathcal{B}$, which necessarily has order $\bar{b}$. Denote by ϕ_1 the τ_α-invariant regular fibre and $\phi_{\bar{b}}$ the τ_α-invariant exceptional fibre. The image in $L(p,q)$ of ϕ_1 is a fibre of multiplicity 1 if ϕ_1 lies in the fixed-point set of τ_α, and is 2 otherwise. Similarly the image in $L(p,q)$ of $\phi_{\bar{b}}$ is a fibre of multiplicity $\bar{b} > 1$ if $\phi_{\bar{b}}$ lies in the fixed-point set of τ_α, and is $2\bar{b}$ otherwise. As $\overline{\mathcal{B}}$ can have only two cone points, it follows that ϕ_1 lies in the fixed point set of τ_α and that its image in $L(p,q)$ is a regular fibre contained in L. Further, $\overline{\mathcal{B}} = S^2(\bar{b}, c)$ if $\phi_{\bar{b}}$ is contained in the fixed-point set of τ_α and $S^2(2\bar{b}, c)$ otherwise. It is easy to see that the lemma holds in this situation. A similar argument shows that the lemma holds if $b = 2$.

Finally suppose that $\bar{\tau}_\alpha$ is not the identity and $a, b > 2$ and define $\mathcal{F}$ to be the fixed point set of $\bar{\tau}_\alpha$. In this case $S^2(\bar{a}, \bar{b}, c, c)$ contains four cone points. Then $\mathcal{F}$ contains either two cone points or none. The former case does not occur as otherwise $\overline{\mathcal{B}}$ would have three cone points (since $a, b \geq 4$). Thus $\mathcal{F}$ contains no cone points and the two associated regular fibres must lie in the fixed point set of τ_α. Thus L contains two regular fibres and $\overline{\mathcal{B}} = S^2(\bar{b}, c)$. This completes the proof in this final case. ♢

For the rest of this section we suppose that $\Delta = \Delta(\alpha, \beta) \geq 5$ and show that this leads to a contradiction.

Recall that L is a 4-braid in the Heegaard solid torus V in $L(p,q)$, and consists of Seifert fibres in the induced Seifert structure on $L(p,q)$. Let X be the exterior of L in $L(p,q)$ and Y the exterior of L in V. Thus X inherits a Seifert structure. By Lemma 7.6(3) at least one component of L is a regular fibre in $L(p,q)$. We distinguish two cases.

CASE I. At least two components of L are regular fibres.

CASE II. Exactly one component of L is a regular fibre.

LEMMA 7.7. *Let K be a component of L that is a regular fibre of $L(p,q)$. Then*

(1) *the winding number of K in V is greater than* 1*;*
(2) *in Case II the winding number of K in V is greater than* 2.

PROOF. Note that because L is a braid in V, K is too. Since the Seifert structure on $L(p,q)$ has two exceptional fibres by Lemma 7.6(2), the exterior of a regular fibre has a Seifert structure with base orbifold $D^2(a,b)$, $a, b \geq 2$. If the winding number of K in V is 1 then the exterior of K in $L(p,q)$ is a solid torus. This proves (1).

To prove (2), suppose K has winding number 2 in V; thus K is a 2-braid in V. Then the exterior of K in $L(p,q)$ has a Seifert structure with base orbifold of the form $D^2(2,c)$. Hence a or $b = 2$, contradicting Lemma 7.6(3). ♢

It follows from Lemma 7.7 that in Case I L has exactly two components, K_1 and K_2, say, each with winding number 2 in V. Let T_1 and T_2 be the corresponding boundary components of X. There is a vertical annulus A in X with boundary components $a_1 \subset T_1$, $a_2 \subset T_2$.

In Case II, let K be the component of L that is a regular fibre. By Lemma 7.7, either $K = L$ or L has two components K and K', with winding numbers 3 and 1, respectively, in V. Let T_K be the boundary component of X corresponding to K. There is a vertical essential annulus A in X, with $\partial A \subset T_K$, separating X into two components X_1 and X_2. Note that either X_1 and X_2 are both solid tori, or one is a solid torus and the other is homeomorphic to $T^2 \times I$.

In both Cases I and II, choose A among all annuli with the stated properties to have minimal intersection with the complementary solid torus to V in $L(p,q)$. We adopt the construction and terminology described in [**BGZ3**, Section 6]. Thus a meridian disk D of V gives a properly embedded 4-punctured disk P in Y, and from A we get an essential n-punctured annulus Q in Y. The intersection of P with Q defines graphs Γ_P, Γ_Q with vertices d_V, c_1, c_2, c_3, c_4 and $a_1, a_2, b_1, \dots, b_n$, respectively. Note that in Case II n is even. Also, since L is a braid we may orient the components of L coherently in V; then the vertices c_1, c_2, c_3, c_4 all have the same sign. By the remarks after Lemma 7.7, each vertex c_j of Γ_P has valency 1 in Case I, and in Case II, valency 0 or 2, with at least three having valency 2.

LEMMA 7.8. *In Case I, Γ_P does not contain a D-edge Scharlemann cycle.*

PROOF. Such a Scharlemann cycle could be used to construct an annulus $A' \subset X$ with $\partial A' = \partial A$ and having fewer intersections with T_V. ◇

LEMMA 7.9. *In Case II, suppose Γ_P contains a D-edge Scharlemann cycle of order $m = 2$ or 3, lying in X_i. Then X_i is a solid torus and the core of X_i is an exceptional fibre of X of multiplicity m.*

PROOF. Let the face of Γ_P bounded by the Scharlemann cycle be f. Note that the edges of ∂f in Γ_Q do not lie in a disk in A, for otherwise X_i would contain a punctured lens space.

Let H be the component of the intersection of the filling solid torus $\overline{L(p,q) - V}$ with X_i corresponding to the label-pair of the Scharlemann cycle. Define $W = N(A \cup H \cup f)$. It is easy to see that W is a solid torus in which A has winding number m. Therefore X_i is not homeomorphic to $T^2 \times I$. Hence X_i is a solid torus, and so $\overline{X_i - W}$ is a solid torus in which the annulus that is the frontier of W in X_i is longitudinal. It follows that the core of X_i is an exceptional fibre of X of multiplicity m. ◇

COROLLARY 7.10. *In both Cases I and II, Γ_P does not contain a D-edge S-cycle.*

PROOF. In Case I this follows from Lemma 7.8. In Case II it follows from Lemma 7.9 together with Lemma 7.6(3). ◇

LEMMA 7.11. *In Case II, Γ_P does not contain D-edge Scharlemann cycles f_1 and f_2 on distinct label-pairs that lie in the same component X_i, where f_j has order 2 or 3, $j = 1, 2$.*

PROOF. Let $H_1, H_2 \subset X_i$ be the components of the intersection of the filling solid torus $\overline{L(p,q) - V}$ with X_i corresponding to the label-pairs of f_1, f_2, respectively. Let $W_1 = N(A \cup H_1 \cup f_1)$. By the proof of Lemma 7.9, W_1 is a solid torus and the annulus A_1 that is the frontier of W_1 in X_i is longitudinal in the solid torus $U = \overline{X_i - W}$. Let $W_2 = N(A_1 \cup H_2' \cup f_2') \subset U$, where $H_2' = H_2 \cap U$ and $f_2' = f_2 \cap U$. Again by the proof of Lemma 7.9, A_1 has winding number 2 or 3 (the order of the Scharlemann cycle f_2) in the solid torus W_2, and hence in U. This is a contradiction. ◇

We will first dispose of Case I, and Case II when $n \geq 4$. Note that the valency of d_V in Γ_P is $\Delta n \geq 5n$ (since $\Delta \geq 5$ by assumption).

For an edge $\bar{e}$ in the reduced graph $\bar{\Gamma}_P$, we denote by $wt(\bar{e})$ the weight of $\bar{e}$, i.e. the number of parallel edges represented by $\bar{e}$. Let $\bar{e}$ be a D-edge of the reduced graph $\bar{\Gamma}_P$. We say $\bar{e}$ is of *type O* if $\bar{e}$ cuts off a subdisk of D that contains a single vertex c_j. Otherwise, $\bar{e}$ is of *type N*. Define

$$\lambda(\bar{e}) = \begin{cases} 2\,wt(\bar{e}), \text{if } \bar{e} \text{ is of type } N; \\ 2\,wt(\bar{e}) + \begin{pmatrix} \text{number of edges of } \Gamma_P \\ \text{incident to the vertex } c_j \end{pmatrix}, \text{if } \bar{e} \text{ of type } O. \end{cases}$$

By Corollary 7.10 and the fact that (by the parity rule) no D-edge can have the same label at both endpoints, it is easy to see that, assuming $n \geq 4$ in Case II, we have $\lambda(\bar{e}) \leq n$ for every D-edge $\bar{e}$ of $\bar{\Gamma}_P$.

Let k be the number of D-edges of $\bar{\Gamma}_P$. If $k = 0$, then $5n \leq 4$ in Case I, and $5n \leq 8$ in Case II, both contradictions. So we assume $k \geq 1$. Also, if $n = 1$ then $k = 0$, so we also assume $n > 1$. Let k_0 be the number of D-edges of $\bar{\Gamma}_P$ of type O. It is easy to see that $k \leq 5$ and $k - 1 \leq k_0 \leq 4$. Note that there are $(4 - k_0)$ vertices c_j that are not associated with D-edges of type O. It follows that

$$5n \leq \Delta n \leq kn + \begin{cases} (4 - k_0)\ , & \text{in Case I;} \\ 2(4 - k_0)\ , & \text{in Case II, if } n \geq 4 \end{cases}$$

LEMMA 7.12. *Case I is impossible.*

PROOF. We have $5n \leq kn + (4 - k_0) \leq kn + (4 - (k-1))$, and hence $5(n-1) \leq k(n-1)$. Since $n > 1$, this implies that $k = 5$ and $k_0 = 4$. Thus $\bar{\Gamma}_P$ has four D-edges of type O and one of type N.

Now if $\bar{e}$ is of type N then

$$\lambda(\bar{e}) \leq \begin{cases} n\ , & n \text{ even;} \\ n - 1\ , & n \text{ odd;} \end{cases}$$

and if $\bar{e}$ is of type O then (since each c_j has valency 1)

$$\lambda(\bar{e}) \leq \begin{cases} n - 1\ , & n \text{ even;} \\ n\ , & n \text{ odd.} \end{cases}$$

Thus we get $5(n-1) < 5(n-1)$, a contradiction. ◇

LEMMA 7.13. *In Case II, $n = 2$.*

PROOF. Assume $n \geq 4$. We have $5n \leq kn + 2(4 - k_0) \leq kn + 2(4 - (k-1))$, giving $(5 - k)n \leq 2(5 - k)$. This is a contradiction unless $\Delta = k = 5$ and $\lambda(\bar{e}) = n$ for each of the five D-edges of $\bar{\Gamma}_P$. The D-edges of $\bar{\Gamma}_P$ are as shown in Figure 7.4. (Each c_j has valency 0 or 2 in Γ_P; the CD-edges of $\bar{\Gamma}_P$ are not shown in Figure 7.4.) Corresponding to the faces $\bar{f}_1, \bar{f}_2$ of $\bar{\Gamma}_P$ shown in Figure 7.4 are 3-gon D-edge faces f_1, f_2 of Γ_P. Since $\lambda(\bar{e}) = n$ for each D-edge $\bar{e}$ of $\bar{\Gamma}_P$, f_1 and f_2 are Scharlemann cycles of order 3.

If $n \equiv 2 \pmod 4$ then f_1 (say) $\subset X_1$ and $f_2 \subset X_2$. Hence each of the two exceptional fibres of $L(p,q)$ has multiplicity 3. But this contradicts Lemma 7.6(2).

If $n \equiv 0 \pmod 4$ then f_1 and f_2 both lie in X_1, say. Since the label-pairs at the corners of f_1 and f_2 are distinct, this contradicts Lemma 7.11. ◇

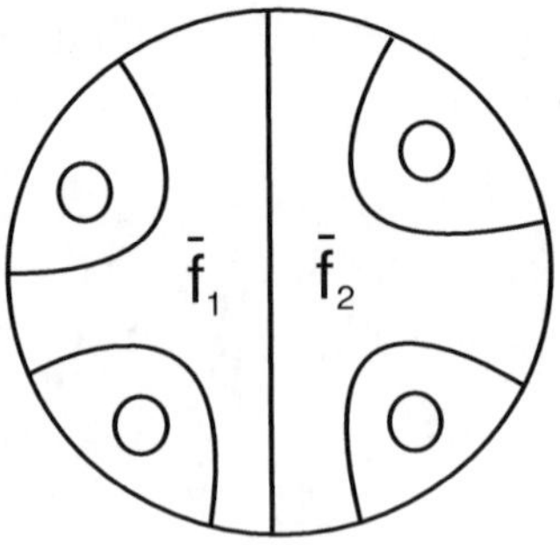

FIGURE 7.4. The D-edges of $\overline{\Gamma}_P$

From now on we will assume that we are in Case II and $n = 2$. Note that there are no parallel D-edges in Γ_P by Corollary 7.10.

Since all the c_j's have the same sign, and a_1, a_2 (resp. b_1, b_2) have opposite signs, we may assume that a_1, a_2 and b_1, b_2 are numbered so that any AB-edge of Γ_Q joins a_i to b_i, $i = 1$ or 2; equivalently, every CD-edge of Γ_P has the same label (1 or 2) at both its endpoints.

Let f be a face of Γ_P. Then ∂f consists of edges of Γ_P together with *corners* at the vertices, i.e. subarcs of the corresponding boundary components of P. We call a corner at the vertex d_V a *d_V-corner*, and a corner at a vertex c_j a *c-corner*.

LEMMA 7.14. *Suppose Γ_P has a disk face with at least two d_V-corners, contained in X_i. Then X_i is a solid torus.*

PROOF. Suppose X_i is not a solid torus; then $X_i \cong T^2 \times I$. Let T_i, T_i' be the boundary components of X_i, where $A \subset T_i$. Let f be a disk face of Γ_P as in the statement of the lemma. Let $H_i = (\overline{L(p,q) - V}) \cap X_i$, and define $W = N(T_i \cup H_i \cup f)$. Then $\partial W = T_i \cup T_i''$, say. Let E be a disk in A containing $H_i \cap A$ and let $x \in H_1(W, E) \cong H_1(W)$ be the class of the core of the 1-handle H_i. Then $H_1(W) \cong (H_1(T_i) \oplus \mathbb{Z})/([\partial f])$, where $\mathbb{Z}$ is generated by x. If f has $m \geq 2$ d_V-corners then $[\partial f] = (y, mx)$ for some $y \in H_1(T_i)$. Therefore $H_1(W, T_i) \cong \mathbb{Z}/m$. But this contradicts the fact that, since T_i'' is homologous to T_i in $X_i \cong T^2 \times I$, $W \cong T^2 \times I$. ♢

Let f be a face of Γ_P. We say f is *of type* (a, b) if f is a disk and has a d_V-corners and b c-corners.

LEMMA 7.15. *Γ_P does not have faces of types (a, b) and (c, d), where $c \geq 2$ and $ad - bc = \pm 1$, that both lie in the same X_i.*

PROOF. Let f, g be faces of type (a, b), (c, d), respectively, lying in X_i. The existence of g implies that X_i is a solid torus by Lemma 7.14.

Let A_i be the annulus $\overline{\partial X_i - A}$. Then $H_1(X_i, A)$ is generated by t, where t is represented by a cocore arc of A_i running from a_2 to a_1. Also, $H_1(X_i, A) \cong \mathbb{Z}/m$, where m is the multiplicity of the exceptional fibre that is the core of X_i. Let $H_i = (\overline{L(p,q) - V}) \cap X_i$, and $\partial_0 H_i = \partial H_i \cap T_V$. The boundary of the face f, as it lies in X_i, consists of a cocore arcs of the annulus $\partial_0 H_i$ (coming from the d_V-corners of f), b cocore arcs of the annulus A_i, (coming from the c-corners of f), together with edges of Γ_Q, lying in A; similarly for g. Let x be the element of $H_1(X_i, A)$

represented by the core of the 1-handle H_i, oriented from b_1 to b_2. Then f and g give the relations

$$ax + bt = 0$$
$$cx + dt = 0$$

in $H_1(X_i, A)$. Since $ad - bc = \pm 1$, this implies that $t = 0$ and therefore $H_1(X_i, A) = 0$, a contradiction. $\diamondsuit$

COROLLARY 7.16. *Γ_P does not have a pair of faces of the following types lying in X_i:*

(1) $(1,1)$ *and* $(2,1)$*;*
(2) $(1,1)$ *and* $(3,2)$*;*
(3) $(2,1)$ *and* $(3,2)$*.*

LEMMA 7.17. *In Γ_Q, the endpoints of the AB-edges incident to b_i are consecutive around b_i, $i = 1, 2$.*

PROOF. This follows immediately from the fact that all the AB-edges incident to b_i have their other endpoint on a_i, $i = 1, 2$. $\diamondsuit$

We will use Lemma 7.17 in conjunction with the following observation. For $i = 1$ or 2, consider the edge-endpoints labeled i around d_V in Γ_P. In Γ_Q these correspond to the edge-endpoints around vertex b_i. Number these $p_0, p_1, \ldots, p_{\Delta-1}$ in order around b_i. Then around d_V the corresponding points with label i occur in the order $p_0, p_d, \ldots, p_{(\Delta-1)d}$, for some integer d coprime to Δ. In particular, if $\Delta = 6$ then $d = \pm 1$. Since AB-edges in Γ_Q correspond to CD-edges in Γ_P we have the following corollary to Lemma 7.17.

COROLLARY 7.18. *If $\Delta = 6$ then the endpoints with label i on d_V of the CD-edges of Γ_P are consecutive among all the edge-endpoints with label i on d_V, $i = 1, 2$.*

LEMMA 7.19. *Case II, $n = 2$, is impossible.*

PROOF. Let k be the number of D-edges of Γ_P, and ℓ the number of vertices c_j with valency 2. Thus $1 \leq k \leq 5$ and $\ell = 3$ or 4.

$\underline{k = 5}$. The D-edges of Γ_P are as shown in Figure 7.5. Then each of X_1 and X_2 contains a D-edge Scharlemann cycle of order 3. By Lemma 7.9 $L(p,q)$ has two exceptional fibres of multiplicity 3. This contradicts Lemma 7.6(2).

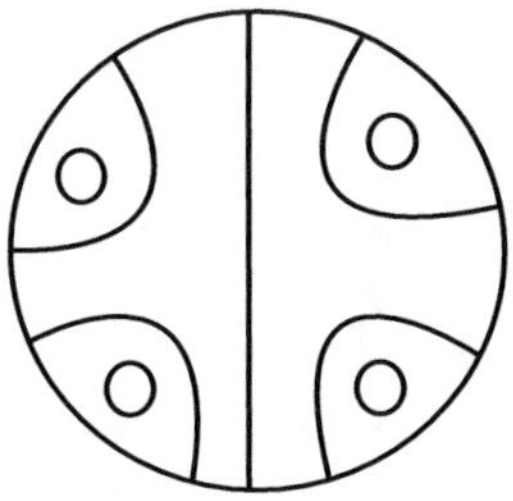

FIGURE 7.5. The D-edges of Γ_P when $k = 5$.

$\underline{k=4}$. There are two cases, (a) and (b), where the D-edges of Γ_P are as shown in Figure 7.6(a) and (b), respectively.

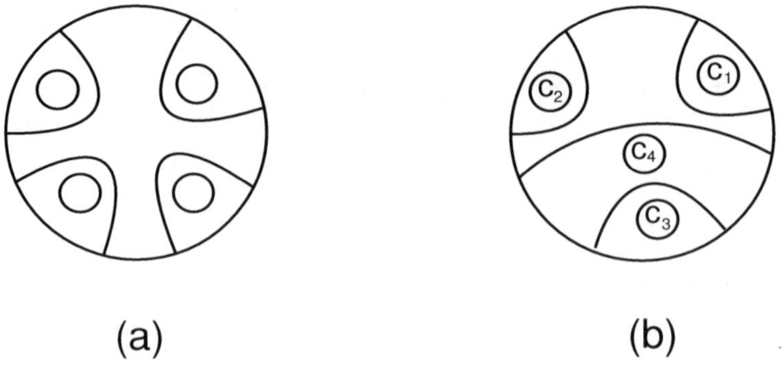

FIGURE 7.6. The two possible arrangements for the D-edges of Γ_P when $k=4$.

CASE (a). A vertex c_j of valency 2 gives rise to a face of type $(2,1)$. It follows from Lemma 7.14 that $\ell = 4$, and hence Γ_P is as shown in Figure 7.7. In particular $\Delta = 8$. Let the edge-endpoints around vertex b_1 in Γ_Q be $p_0, p_1, \ldots, p_7$, numbered in order around the vertex. By Lemma 7.17 we may assume that p_0, p_1, p_2, p_3 are endpoints of AB-edges and p_4, p_5, p_6, p_7 endpoints of B-edges. By the remarks after Lemma 7.17 the corresponding points with label 1 on d_V in Γ_P appear in the order $p_0, p_d, \ldots, p_{7d}$ for some d coprime to 8. Since we see from Figure 4 that the CD-edges and D-edges with label 1 alternate around d_V it is clear that no such integer d exists.

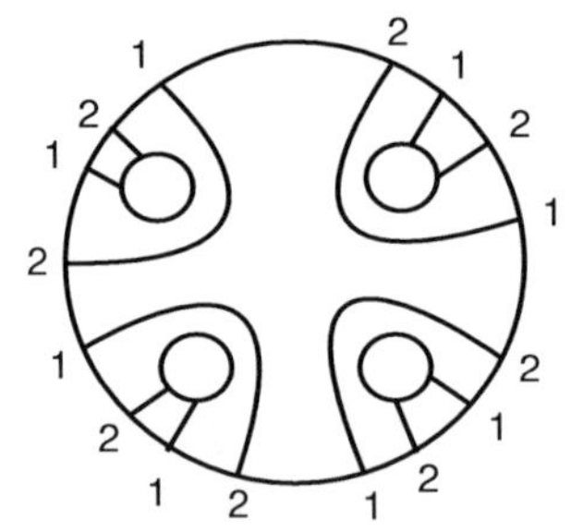

FIGURE 7.7. Γ_P in Case (a) of $k=4$

CASE (b). First note that at least one of c_1, c_2 has valency 2, and hence both do by Lemma 7.14. Similarly, Lemma 7.14 implies that c_4 has valency 2. There are two possibilities for the edges incident to c_4, shown in Figure 7.8(i) and (ii).

In (i) the faces f and g contradict Corollary 7.16(1).

In (ii), the face f implies, by Lemma 7.14, that c_3 has valency 2. Hence there is a face of type (1,1) on the same side as the face g, contradicting Corollary 7.16(1). (Here, and in the sequel, by "on the same side" we shall mean on the same side of A in X, i.e. in the same component X_1 or X_2.)

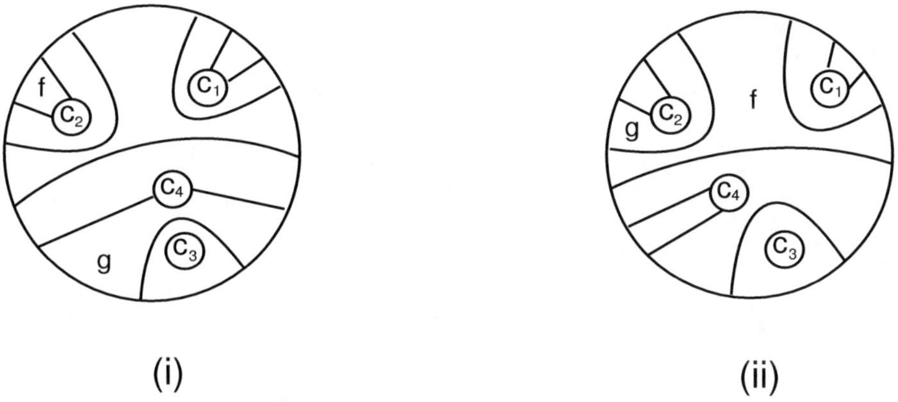

FIGURE 7.8. The two possible arrangements for the edges incident to c_4 in Case (b) of $k = 4$

$\underline{k = 3}$. There are three cases, (a), (b) and (c), where the D-edges of Γ_P are as shown in Figure 7.9(a), (b) and (c), respectively.

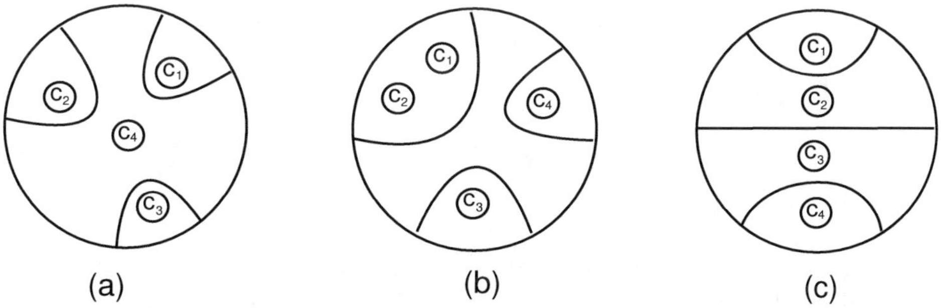

FIGURE 7.9. The three possible arrangements for the D-edges of Γ_P when $k = 3$

CASE (a). If c_4 has valency 0 then Γ_P is as shown in Figure 7.10. Hence $\Delta = 6$. Since the 1-labels on d_V belonging to CD-edges are not consecutive around d_V, this contradicts Corollary 7.18.

So suppose c_4 has valency 2. The two possibilities for the edges incident to c_4 are shown in Figure 7.11(i) and (ii).

In (i), the face f implies that c_1 has valency 2, by Lemma 7.14. Hence there is a face g of type (2,1) on the same side as f. Also, at least one of c_2, c_3 has valency 2; this gives a face of type (1,1) on the same side as g, contradicting Corollary 7.16(1).

In subcase (ii), since some c_j, $j = 1, 2, 3$, has valency 2, we get a face of type (2,1) on the same side as f, contradicting Corollary 7.16(1).

CASE (b). At least one of c_3, c_4 has valency 2, and hence all the c_j's have valency 2 by Lemma 7.14. The four possibilities for Γ_P are shown in Figure 7.12(i), (ii), (iii) and (iv).

(i) contradicts Corollary 7.16(3), and (ii) contradicts Corollary 7.16(1).

In (iii), $\Delta = 6$ and the labeling around d_V contradicts Corollary 7.18.

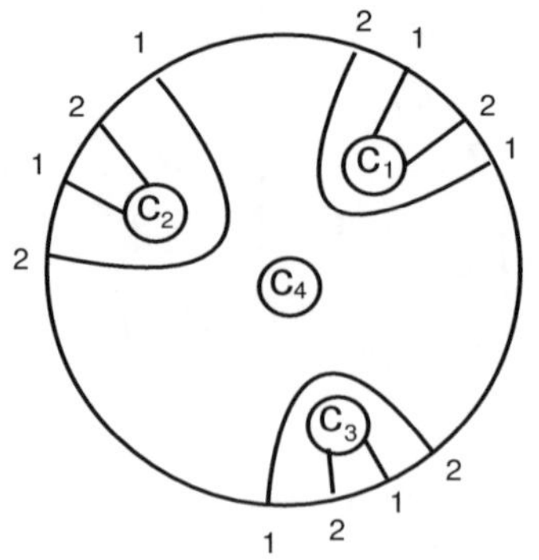

FIGURE 7.10. Γ_P when c_4 has valency 0 in Case (a) of $k = 3$

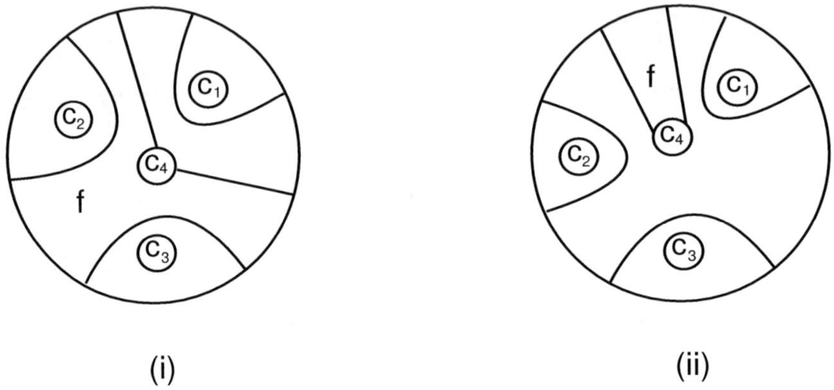

FIGURE 7.11. The two possible arrangements for the two edges incident to c_4 when c_4 has valency 2 in case (a) of $k = 3$.

In (iv), suppose the face f lies in X_1. Then X_1 is a solid torus by Lemma 7.14, and f gives the relation $2x + t = 0$ in $H_1(X_1, A)$. (See the proof of Lemma 7.15 for notation.) The annulus face g also lies in X_1 and gives the relation $x + 2t = 0$. These relations give $3t = 0$, and hence $H_1(X_1, A) \cong \mathbb{Z}/3$. This implies that the core of X_1 is an exceptional fibre of multiplicity 3. The D-edge Scharlemann cycle h implies that the core of X_2 is also an exceptional fibre of multiplicity 3, by Lemma 7.9. This contradicts Lemma 7.6(2).

Case (c). Since $\ell \geq 3$, either c_1 and c_2 both have valency 2, or c_3 and c_4 both have valency 2. Assume the former without loss of generality. The fact that c_1 has valency 2 gives a face f of type (1,1) and a face g of type (2,1) on opposite sides. There are two possibilities for the configuration of the CD-edges incident to c_2. In one case we get a face of type (1,1) on the same side as g, and in the other case a face of type (2,1) on the same side as f. These both contradict Corollary 4.16(1).

$\underline{k = 2}$. There are two cases (a) and (b), illustrated in Figure 7.13(a) and (b).

CASE (a). First suppose that c_3 has valency 0. Then c_1, c_2 and c_4 have valency 2, and since $\Delta \geq 5$ the edges incident to c_1, c_2 and c_4 are all CD-edges. This gives faces of type (2,1) and (3,2) on the same side, contradicting Corollary 7.16(3).

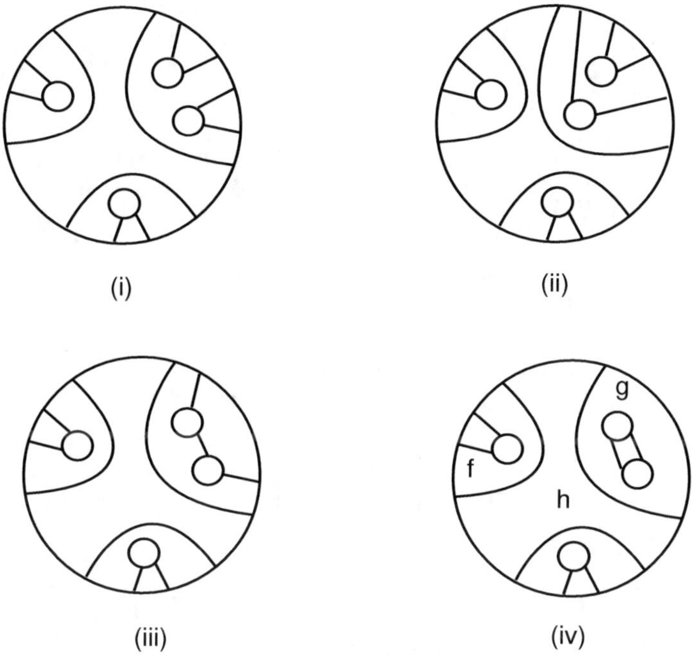

FIGURE 7.12. The four possibilities for Γ_P when all the c_j's have valency 2 in Case (b) of $k = 3$.

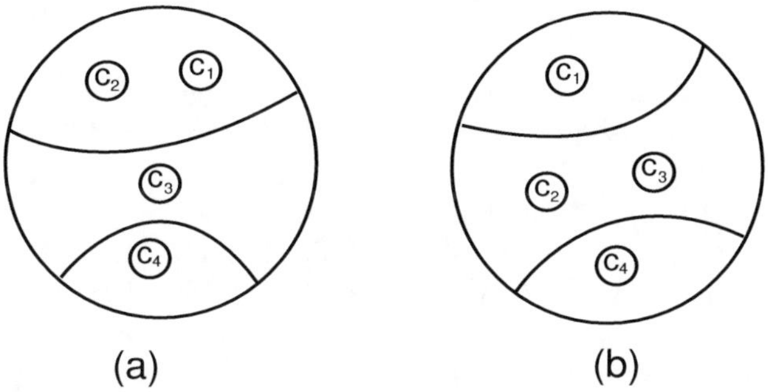

FIGURE 7.13. The two possibilities for the D-edges of Γ_P when $k = 2$.

So suppose c_3 has valency 2. The two possible arrangements of the edges incident to c_3 are shown in Figure 7.14(i) and (ii).

In (i), the face f, together with Lemma 7.14, implies that c_4 has valency 2. This gives a face of type (1,1) on the same side as g, contradicting Corollary 7.16(1).

In subcase (ii), if c_4 has valency 2 then we get a face of type (2,1) on the same side as f, contradicting Corollary 7.16(1). If c_4 has valency 0 then c_1 and c_2 have valency 2, and since $\Delta \geq 5$ the edges incident to c_1 and c_2 are CD-edges. The resulting face of type (3,2) contradicts Lemma 7.14.

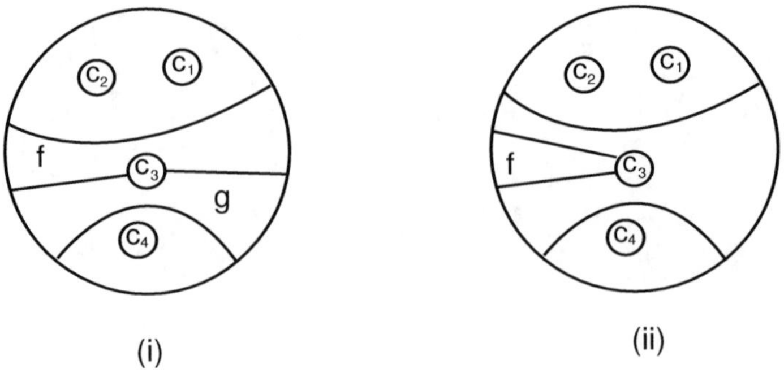

FIGURE 7.14. The two possibilities for the edges incident to c_3 when c_3 has valency 2 in Case (a) of $k = 2$.

CASE (b). First suppose that one of c_2, c_3 has valency 0. Note that since $\Delta \geq 5$ Γ_P has no C-edges. Therefore there are three possibilities for Γ_P, illustrated in Figure 7.15(i), (ii) and (iii).

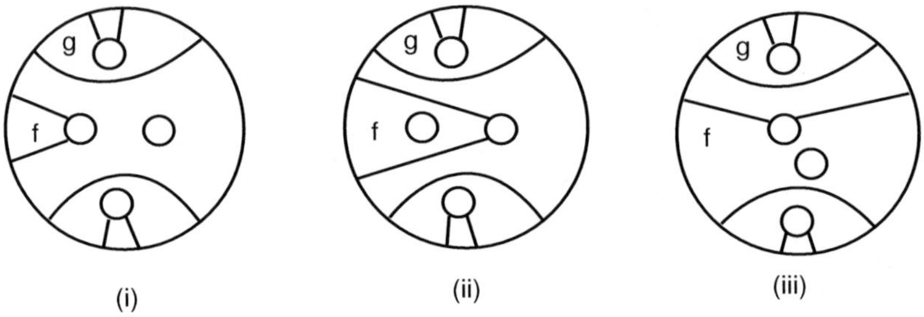

FIGURE 7.15. The three possibilities for Γ_P when one of c_2, c_3 has valency 0 in Case (b) of $k = 2$.

In (i) the faces f and g contradict Corollary 7.16(1).

In (ii) and (iii) the faces f and g contradict Lemma 7.14.

So suppose that both c_2 and c_3 have valency 2. Since $\Delta \geq 5$ there cannot be two C-edges joining c_2 and c_3. There are therefore seven possibilities for the edges incident to c_2 and c_3, shown in Figure 7.16(i)–(vii).

In (i) and (ii), at least one of c_1 and c_4 has valency 2. This gives a face of type (2,1) on the same side as f, contradicting Corollary 7.16(1).

In (iii), the face f implies, by Lemma 7.14, that both c_1 and c_4 have valency 2. Hence Γ_P is as illustrated in Figure 7.17. In particular $\Delta = 6$. But the 1-labels on d_V of the CD-edges are not consecutive on d_V, contradicting Corollary 7.18.

In (iv), the faces f and g contradict Corollary 7.16(1).

In (v), at least one of c_1 and c_4 has valency 2, giving a face of type (1,1) on the same side as f. This contradicts Corollary 7.16(1).

In (vi) and (vii) we must have $\ell = 4$ since $\Delta \geq 5$. Then in (vi) we get faces of type (1,1) and (3,2) on the same side, and in (vii) faces of type (1,1) and (2,1) on the same side, contradicting Corollary 7.16(2) and (1) respectively.

<u>$k = 1$</u>. The two possibilities for the D-edge of Γ_P are shown in Figure 7.18 (a) and (b). Note that since $\Delta \geq 5$, Γ_P has no C-edges and $\ell = 4$.

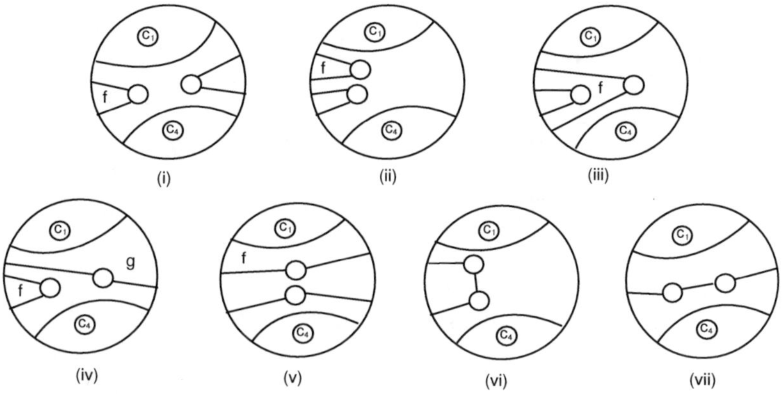

FIGURE 7.16. The seven possibilities for the edges incident to c_2 and c_3 when both of then have valency 2 in Case (b) of $k = 2$.

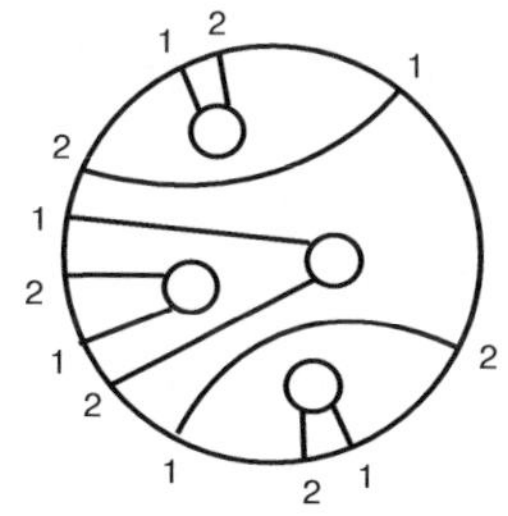

FIGURE 7.17. Γ_P in Case (iii) of Figure 7.16

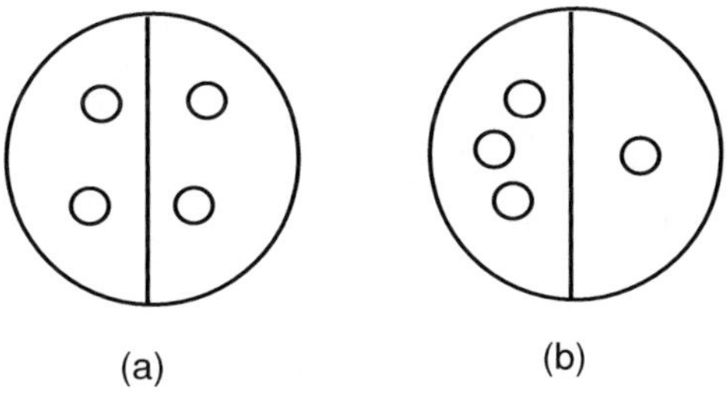

FIGURE 7.18. The two possibilities for the D-edge of Γ_P when $k = 1$.

CASE (a). Either Γ_P is as shown in Figure 7.19(i), or the CD-edges on one side of the D-edge are as shown in Figure 7.19(ii).

In (i) we have faces of type (1,1) and (3,2) on the same side, contradicting Corollary 7.16(2).

In (ii) we have faces of type (1,1) and (2,1) on the same side, contradicting Corollary 7.16(1).

CASE (b). It is easy to see that for all possible configurations of the CD-edges we get faces of type (1,1) and (2,1) on the same side, contradicting Corollary 7.16(1).

This completes the proof of Lemma 7.19. ◇

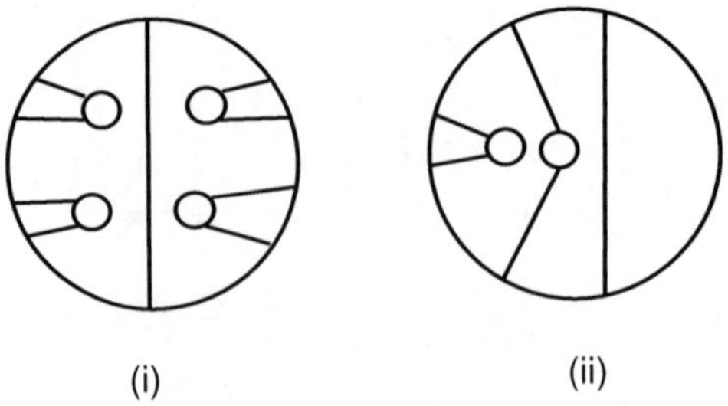

FIGURE 7.19. The two possibilities for Γ_P in Case (a) of $k = 1$.

The not very small case of Proposition 7.1 now follows from Lemmas 7.12, 7.13 and 7.19. $\diamondsuit$

CHAPTER 8

The proof of Theorem 1.2 when F is a fibre

In this chapter we prove the fibre case of Theorem 1.2.

PROPOSITION 8.1. *Suppose that M is a hyperbolic knot manifold which contains an essential twice-punctured torus F of boundary slope β and let α be a slope on ∂M such that $M(\alpha)$ is an irreducible small Seifert manifold. If F is a fibre in M, then $\Delta(\alpha, \beta) \leq 3$.*

Recall the involution τ_F depicted in Figure 7.1 and let τ be its fibre-by-fibre extension to M. Then $M/\tau = (V, L)$ where V is a solid torus and L is a 4-braid. The reader will verify that there is curve representing a dual class β^* to β on ∂M on which τ acts by rotation by π. Write

$$\alpha = p\beta^* + q\beta \tag{8.1}$$

where p, q are coprime. After possibly changing the signs of β^* and β we may assume that

$$p = \Delta(\alpha, \beta) \tag{8.2}$$

Note that β is sent to the meridian slope $\bar{\beta}$ on ∂V, so $V(\bar{\beta}) \cong S^1 \times S^2$, while β^* is sent to $2\bar{\beta}^*$ where $\bar{\beta}^*$ is a longitude of V.

For each slope γ on ∂M, τ extends to an involution $\tau_\gamma : M(\gamma) \to M(\gamma)$. Moreover, if $\widetilde{U}_\gamma$ denotes the filling torus in $M(\gamma)$ and $\widetilde{K}_\gamma$ its core, then

$$Fix(\tau_\gamma) = \begin{cases} \widetilde{L} & \text{if } \Delta(\gamma, \beta^*) \text{ is odd} \\ \widetilde{L} \cup \widetilde{K}_\gamma & \text{if } \Delta(\gamma, \beta^*) \text{ is even} \end{cases} \tag{8.3}$$

It is clear that $\widetilde{U}_\gamma/\tau_\gamma$ is a solid torus U_γ. Denote its core $\widetilde{K}_\gamma/\tau_\gamma$ by K_γ. Thus $M(\gamma)/\tau_\gamma = V \cup_{\bar{\gamma}} U_\gamma$ is a lens space. Indeed, if $\gamma = r\beta^* + s\beta$, then under the double cover $\partial M \to \partial V$ we have $\gamma \mapsto 2r\bar{\beta}^* + s\bar{\beta}$. Let $\bar{\gamma} = \frac{1}{\gcd(2,s)}(2r\bar{\beta}^* + s\bar{\beta})$ denote the associated slope and L_γ the branch set in $M(\gamma)/\tau_\gamma$. Then

$$(M(\gamma)/\tau_\gamma, L_\gamma) = (V(\bar{\gamma}), L_\gamma) \cong \begin{cases} (L(2r, s), L) & \text{if } s \text{ is odd} \\ (L(r, \frac{s}{2}), L \cup K_\gamma) & \text{if } s \text{ is even} \end{cases}$$

We are interested in the case $\gamma = \alpha$. Set

$$\bar{p} = \frac{2p}{\gcd(2, q)} \quad \text{and} \quad \bar{q} = \frac{q}{\gcd(2, q)} \tag{8.4}$$

so that $\bar{\alpha} = \bar{p}\bar{\beta}^* + \bar{q}\bar{\beta}$ and

$$M(\alpha)/\tau_\alpha \cong L(\bar{p}, \bar{q})$$

From 8.3 we see that

$$|L_\alpha| = \begin{cases} |L| & \text{if } q \text{ is odd} \\ |L| + 1 & \text{if } q \text{ is even} \end{cases} \tag{8.5}$$

By [**BGZ3**, Lemma 4.1] there is a τ_α-invariant Seifert structure on $M(\alpha)$ with base orbifold of the form $S^2(a,b,c)$ where $a,b,c \geq 1$. Let $\bar{\tau}_\alpha$ be the induced map on $S^2(a,b,c)$.

We begin with the case where $M(\alpha)$ is a lens space.

LEMMA 8.2. *Suppose that M is a hyperbolic knot manifold which contains an essential twice-punctured torus F of boundary slope β and let α be a slope on ∂M such that $M(\alpha)$ is an irreducible small Seifert manifold. If F is a fibre in M and $M(\alpha)$ is a lens space, then $\Delta(\alpha,\beta) \leq 3$.*

PROOF. By [**L3**] $\Delta(\alpha,\beta) \leq 4$ so we suppose that $p = \Delta(\alpha,\beta) = 4$ in order to derive a contradiction. In this case q is odd, so that $L_\alpha = L$ (8.5). Further, $\bar{p} = 8$ and $\bar{q} = q$ so $M(\alpha)/\tau_\alpha \cong L(8,q)$. By [**BGZ3**, Lemma 4.2], L_α is either

- the union of the cores of the two Heegaard solid tori of $L(8,q)$; or
- the boundary of an annular spine of a Heegaard solid torus of $L(8,q)$.

It follows that each component of L_α carries a generator of $H_1(L(8,q))$. On the other hand, L_α is isotopic in $L(8,q)$ to the 4-braid L in V so L must split into a 1-braid L_0 and a 3-braid L_1. Since the generators of $H_1(L(8,q))$ have order 8, L_0 and L_1 carry different generators and so L_α must be the union of the cores of the two Heegaard solid tori of $L(8,q)$.

CLAIM 1. Let Y_1 be the exterior L_1 in V. Then Y_1 is Seifert fibred.

The Dehn filling of Y_1 along ∂V with the slope $\bar{\alpha}$ is a manifold homeomorphic to the exterior of L_1 in $L(8,q)$ and thus is a solid torus. On the other hand the Dehn filling of Y_1 along ∂V with the slope $\bar{\beta}$ is a manifold homeomorphic the exterior of L_1 in $S^2 \times S^1$ and thus is a twice-punctured disk bundle and therefore Seifert fibred. Since $\Delta(\bar{\alpha},\bar{\beta}) = 8 > 5$, Y_1 cannot be hyperbolic by [**Go1**, Theorem 1.3]. Since Y_1 is the exterior of a single component closed 3-braid in V, the monodromy of the corresponding thrice-punctured disk bundle structure on Y_1 cannot be reducible. So it is periodic and Y_1 is Seifert fibred.

CLAIM 2. Let Y be the exterior of $L = L_0 \cup L_1$ in V. Then Y is hyperbolic.

To prove this claim, suppose for contradiction that Y is non-hyperbolic. It is irreducible, since L_0 and L_1 are each homotopically essential in V. Thus Y is either Seifert fibred or contains an essential torus. In the latter case, the monodromy of Y with respect to the 4-punctured disk bundle structure is reducible and its invariant essential curve in a 4-punctured disk fiber can only be a circle which encloses the three punctures coming from L_1 but not the puncture from L_0. Hence the corresponding essential torus T separates Y into a thrice-punctured disk bundle Y_1 and a once-punctured annulus bundle Y_0. Obviously Y_0 is Seifert fibred and Y_1 is homeomorphic to the exterior of L_1 in V, which is Seifert fibred by Claim 1. Therefore Y has no hyperbolic pieces in its JSJ torus decomposition. Now the exterior of $\tilde{L}$ in M is a free double cover of Y and thus has no hyperbolic pieces in its JSJ torus decomposition. Thus M cannot be hyperbolic, giving a contradiction.

To finish the proof of the lemma, we adopt an intersection graph argument as in §7.2. In the present case the argument is much simpler due to the relatively large distance of 8 between the slopes $\bar{\alpha}$ and $\bar{\beta}$ on ∂V.

Let T_i be the torus boundary of Y corresponding to L_i, $i = 0,1$. Then $\partial Y = T_0 \cup T_1 \cup \partial V$. The $\bar{\alpha}$-filling of Y along ∂V is the exterior of L in $L(8,q)$ and thus is homeomorphic to $T^2 \times [0,1]$ (recall that L_0 and L_1 are cores of the two

Heegaard solid tori of $L(8, q)$). Let A be a properly embedded essential annulus in this manifold whose boundary component a_1 on T_1 is a meridional slope of L_1. Then the other boundary component a_0 of A on T_0 is a slope distance 8 from the meridional slope of L_0. Now isotope A fixing ∂A so that $A \cap \partial V$ has minimal number of components and let $Q = Y \cap A$. Then Q is an essential n-punctured annulus in Y. Since Y is hyperbolic, $n > 0$. Let $b_1, \cdots, b_n$ be the components of ∂Q which lie on ∂V and they are indexed in the order they appear along ∂V. Note that $b_1, \ldots, b_n$ are parallel circles in ∂V with the slope $\bar{\alpha}$.

Let P be a fixed 4-punctured disk fiber in Y. Then P is essential in Y. Denote the components of ∂P by d, c_0, c_1, c_2, c_3, where d lies on ∂V and has slope $\bar{\beta}$, c_0 lies on T_0 and has the meridional slope of L_0, and c_1, c_2, c_3 lie on T_1 and each has the meridional slope of L_1.

Isotope Q and P to intersect transversely and minimally, and consider their intersection graphs Γ_Q, Γ_P with vertices $a_0, a_1, b_1, \cdots b_n$ and d, c_0, c_1, c_2, c_3 respectively, and with the arc components of $Q \cap P$ as edges. Since a_1 and c_1, c_2, c_3 are meridians of L_1 on T_1, they are disjoint and thus have valency 0. By construction a_0 and c_0 have valency 8, d has valency $8n$, and $b_1, \cdots, b_n$ have valency 8.

The edges of Γ_Q and Γ_P are essential arcs in P and Q. By the parity rule, no edge in Γ_Q has both of its endpoints incident to a_0 and thus there are exactly 8 edges in Γ_Q incident to a_0, each connecting a_0 to some b_i. Call such edges *ab-edges*. The remaining $4(n-1)$ edges Γ_Q connect b_i's to b_i's and we call them *bb-edges*.

Similarly no edge in Γ_P has both of its endpoints incident to c_0 and thus in Γ_P there are exactly 8 *cd-edges* connecting c_0 to d. The remaining $4(n-1)$ edges of Γ_P connect d to d which are called *dd-edges*.

An edge in one graph is dual to an edge in the other graph if they come from the same arc component of $P \cap Q$. Then *ab*-edges are dual to *cd*-edges and *bb*-edges are dual to *dd*-edges. The parity rule shows that each *dd*-edge is a positive edge and each *bb*-edge is negative (i.e. it connects vertices of opposite signs).

As in Lemma 7.8, Γ_P has no Scharlemann cycles consisting of *dd*-edges. It is an elementary fact that if Γ_P has a parallel family of more than $n/2$ *dd*-edges, then it has a Scharlemann cycle of *dd*-edges (an S-cycle to be exact; see [**CGLS**, Corollary 2.6.7]). Hence each parallel family of *dd*-edges has at most $n/2$ members. It's easy to see that the reduced graph of Γ_P has at most 5 *dd*-edges (cf. Figure 7.5). Hence in Γ_P there are at most $5n/2$ *dd*-edges. On the other hand there are exactly $4(n-1)$ *dd*-edges. Hence we have $4(n-1) \leq 5n/2$, which yields $n \leq 2$.

CLAIM 3. If e_1 and e_2 are two parallel *ab*-edges in Γ_Q, then their dual *cd*-edges e_1' and e_2' cannot be parallel in Γ_P.

To prove the claim assume, for a contradiction, that e_1' and e_2' are parallel in Γ_P. Let B be the disk region in Q between e_1 and e_2 and let B' be the disk region in P between e_1' and e_2'. We may assume that e_1 and e_2 are an innermost such pair, i.e. B and B' only intersect in Y along $e_1 = e_1'$ and $e_2 = e_2'$. (Otherwise there is an edge $e \neq e_1, e_2$ in B whose dual edge e' is contained B', and we may replace e_2 by e and restart the proof.) Then B and B' paste together in Y forming a properly embedded annulus E in Y connecting ∂V and T_0. We claim that E is essential in Y, contrary to Claim 2, thus completing the proof of Claim 3.

Since E connects different components of ∂Y, it suffices to show that each component of ∂E is essential in T_0 or ∂V. To see this, suppose that e_1 and e_2 are incident at b_i. Then the four endpoints of e_1, e_2 decompose ∂B into four arcs: e_1,

a, e_2, b, where a is a sub-arc of a_0 and b a sub-arc of b_i. Similarly the endpoints of e_1', e_2' decompose $\partial B'$ into four arcs: e_1', c', e_2', d', where c' is a sub-arc of c_0 and d' a sub-arc of d. Since a_0 and c_0 intersect minimally in T_0, the arcs a and c' paste together in T_0 to form a simple essential circle. Likewise b and d' paste together in ∂V to form a simple essential circle. Thus E is essential, contrary to Claim 2.

CASE $n = 1$. In this case there are exactly eight edges in Γ_P, all being cd-edges, and there are exactly eight edges in Γ_Q, all being ab-edges. It is clear that the eight ab-edges in Γ_Q are parallel to each other, and it is also evident that there are at least two parallel cd-edges in Γ_P, contrary to Claim 3. Thus $n \neq 1$.

CASE $n = 2$. In this case there are exactly twelve edges in Γ_P, eight being cd-edges and four being dd-edges, and similarly there are exactly twelve edges in Γ_Q, eight being ab-edges and four being bb-edges. Since $n = 2$, the four bb-edges all connect b_1 to b_2. It follows that four of the eight ab-edges in Γ_Q are incident to b_1 and four are incident to b_2. Thus the reduced graph of Γ_Q is one of the three cases (1) (2) (3) shown in Figure 8.1. It follows that there is at least one family of four parallel ab-edges in Γ_Q, which we denote by e_1, e_2, e_3, e_4. Their dual edges, e_1', e_2', e_3', e_4', are cd-edges and it is easy to see that as the valency of c_0 in the reduced graph of Γ_P is at most 3, at least two of them are parallel in Γ_P. Again we obtain a contradiction to Claim 3 and this rules out the final possibility that $n = 2$. ♢

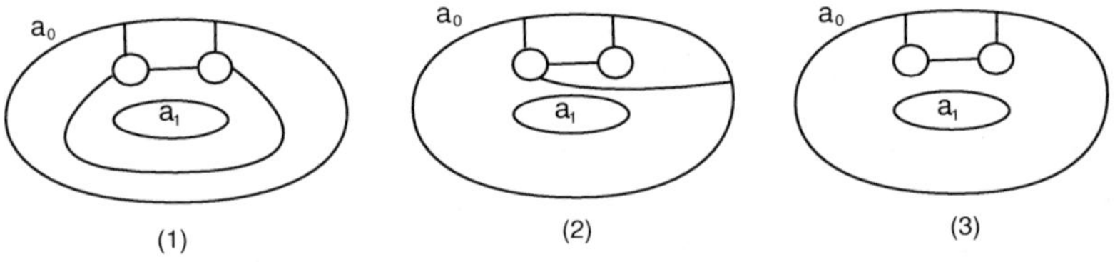

FIGURE 8.1. The three possibilities for the reduced graph of Γ_Q when $n = 2$.

LEMMA 8.3. *Suppose that M is a hyperbolic knot manifold which contains an essential twice-punctured torus F of boundary slope β and let α be a slope on ∂M such that $M(\alpha)$ is an irreducible small Seifert manifold. If F is a fibre in M, $M(\alpha)$ is not a lens space, and τ_α reverses the orientations of the Seifert fibres of $M(\alpha)$, then $\Delta(\alpha, \beta) \leq 3$.*

PROOF. Suppose that $p = \Delta(\alpha, \beta) \geq 3$. Then $\bar{p} \geq 3$, so we can argue as in the proof of [**BGZ3**, Lemma 4.4] to see that $M(\alpha)$ has base orbifold $S^2(\bar{p}, \bar{p}, m)$ where $m \geq 2$. Further, there is an integer n coprime with m such that L_α is isotopic to the closure of an m/n rational tangle in a genus 1 Heegaard solid torus V of $M(\alpha)/\tau_\alpha$ as depicted in Figure 7.2. We also have $|L_\alpha| = 1$ if n is odd and 2 if it is even.

Suppose that m is even. Then L_α is connected, so $L = L_\alpha$ and therefore q is odd (8.5). Further L is homotopically trivial in $L(\bar{p}, \bar{q})$; in fact, L is homotopically trivial in the solid torus V since it is the closure in V of an m/n-rational tangle with m even. But the 4-braid L represents four times a generator of $H_1(L(\bar{p}, \bar{q}))$, so $2p = \bar{p} > 1$ divides 4, which is impossible since we have assumed $p \geq 3$. Thus m is odd, and so is at least 3.

Considering the universal cover of the lens space $L(\bar{p}, \bar{q})$, we obtain two links $\widehat{L}$ and $\widehat{L}_\alpha$ in S^3 where $\widehat{L}$, the lift of the image of L in $L(\bar{p}, \bar{q})$, is a 4-braid and $\widehat{L}_\alpha$, the lift of L_α, is a Montesinos link with $\bar{p}$ rational tangles each of type m/n.

The 2-fold cover $\Sigma_2(\widehat{L}_\alpha)$ of S^3 branched over $\widehat{L}_\alpha$ is Seifert with base orbifold a 2-sphere with $\bar{p}$ cone points, each of order $m \geq 3$. Thus the Heegard genus of $\Sigma_2(\widehat{L}_\alpha))$ is $\bar{p} - 1$ [**BoiZie**]. (Note that any irreducible horizontal Heegaard surface of the Seifert manifold, if one exists, has genus larger than $\bar{p} - 1$.)

If q is odd, then $L_\alpha = L$, so $\widehat{L}_\alpha$ is a closed 4-braid. Since a 2-sphere in S^3 which separates $(S^3, \widehat{L}_\alpha)$ into two trivial 4-string tangles lifts to a genus 3 Heegaard surface in $\Sigma_2(\widehat{L}_\alpha)$, the Heegaard genus of the latter is at most 3. Thus as $\bar{p} = 2p$ when q is odd, $2p - 1 = \bar{p} - 1 \leq 3$. But then $\Delta(\alpha, \beta) = p \leq 2$, contrary to our hypotheses.

Thus q is even, so p is odd and $\bar{p} = p$. In this case, L_α is the union of L and the core of the filling torus in $L(\bar{p}, \bar{q}) = V(\bar{\alpha})$. Hence there is a genus 1 Heegaard splitting $U_1 \cup U_2$ of S^3 such that $\widehat{L}_\alpha$ is the union of the closed 4-braid $\widehat{L} \subset U_1$ and the core of U_2. It follows that $\widehat{L}_\alpha$ is the closure of a 5-braid in S^3, so the Heegaard genus of $\Sigma_2(\widehat{L}_\alpha)$ is at most 4. Hence $p - 1 = \bar{p} - 1 \leq 4$, which gives $p \leq 5$.

Since $p \geq 3$ is odd, $p = 3$ or 5. We will now eliminate the second case.

Assume $p = 5$. Then L is a 4-braid in the Heegaard solid torus $V \subset L(5, \bar{q}) = V(\bar{\alpha})$, and L_α is the union of L and a core of the complementary solid torus. Also, L_α is the closure of the m/n-rational tangle (m odd, n even) in some Heegaard solid torus W in $L(5, \bar{q})$, as shown in Figure 7.2. It follows that each component of L_α is a core of W. In particular, the exterior of L in $L(5, \bar{q})$ is a solid torus.

Let Y be the exterior of L in V. Then ∂Y has two components, ∂V and T_L, say. A meridian disk D of V gives rise to a 4-punctured disk $P \subset Y$, with $\partial P = \partial D \cup c_1 \cup c_2 \cup c_3 \cup c_4$, where the c_j's are meridians of L. Note that ∂D has slope $\bar{\beta}$ on ∂V. Also, since $Y(\bar{\alpha})$, the exterior of L in $L(5, \bar{q})$, is a solid torus, there is an essential disk $E \subset Y(\bar{\alpha})$ with $\partial E \subset T_L$. Choosing E to have minimal intersection with the core of the filling solid torus in $Y(\bar{\alpha})$ we get an essential punctured disk $Q \subset Y$, with $\partial Q = \partial E \cup b_1 \cup \cdots \cup b_n$, where the b_i's are curves of slope $\bar{\alpha}$ on ∂V. Since Y is hyperbolic, $n \geq 2$.

Let ε be the slope of ∂E on T_L, and let μ be the slope on T_L of a meridian of L.

CLAIM 8.4. $\Delta(\varepsilon, \mu) = 5$.

PROOF OF CLAIM 8.4. Let $\lambda, \bar{\gamma}$ be slopes on T_L, ∂V, respectively, such that $\Delta(\lambda, \mu) = 1$, $\Delta(\bar{\gamma}, \bar{\beta}) = 1$. Then $H_1(Y) \cong \mathbb{Z} \oplus \mathbb{Z}$, generated by μ and $\bar{\gamma}$. Also, in $H_1(Y)$ we have $\bar{\beta} = 4\mu$ and $\lambda = 4\bar{\gamma} + r\mu$ for some $r \in \mathbb{Z}$. Write $\varepsilon = a\lambda + b\mu$, where $(a, b) = 1$. Then $\varepsilon = a(4\bar{\gamma} + r\mu) + b\mu = 4a\bar{\gamma} + (b + ra)\mu \in H_1(Y)$. Note that $\langle \varepsilon \rangle = \ker(H_1(T_L) \to H_1(Y(\bar{\alpha})))$. Since $\bar{\alpha} = 5\bar{\gamma} + \bar{q}\bar{\beta} = 5\bar{\gamma} + 4\bar{q}\mu$, it follows that there exists $m \in \mathbb{Z}$ such that

$$4a = 5m \ , \quad \text{and}$$
$$b + ra = 4\bar{q}m \ .$$

The first equation implies $a = 5c$, $m = 4c$ for some $c \in \mathbb{Z}$, and the second then implies $b \equiv 0 \pmod{c}$. Since $(a, b) = 1$, $c = \pm 1$. Hence $\Delta(\varepsilon, \mu) = |a| = 5$. $\diamondsuit$

The intersection of P and Q defines in the usual way graphs Γ_P, Γ_Q in the 2-sphere, where Γ_P has vertices c_1, c_2, c_3, c_4 and d_V (corresponding to ∂D), and Γ_Q has vertices $b_1, b_2, \ldots, b_n$ and e_L (corresponding to ∂E). Thus in Γ_P the vertex d_V has valency $5n$ and each c_j has valency 5, while in Γ_Q the vertex e_L has valency 20 and each b_i has valency 5. Since L is a braid, all the c_j's have the same sign. Hence, by the parity rule, Γ_P has no C-edges, Γ_Q has no E-edges, and all B-edges of Γ_Q are negative.

The computation in the proof of Claim 8.4 shows that $m = \pm 4$ i.e. $\varepsilon = \pm 4\bar{\alpha} \in H_1(Y)$. Thus (with any choice of orientations)

$$\Big| \text{ number of positive } b_i\text{'s} - \text{number of negative } b_i\text{'s } \Big| = 4$$

It follows that n is even and ≥ 4.

CLAIM 8.5. $n = 4$.

REMARK 8.6. This is equivalent to saying that Γ_P has no D-edges.

PROOF OF CLAIM 8.5. Suppose $n \geq 6$. We use the notation and terminology of the argument immediately preceding Lemma 7.12.

Since each c_j has valency 5, it is easy to see that if $\bar{e}$ is an edge of $\bar{\Gamma}_P$ of type O then $\lambda(\bar{e}) \leq n - 1$. If $\bar{e}$ is of type N then $\lambda(\bar{e}) \leq n$. Since the valency of d_V is $5n$ we get

$$5n \leq k_0(n-1) + (k - k_0)n + (4 - k_0)5$$

Since $k - k_0 \leq 1$ this gives

$$(4 - k_0)n \leq 20 - 6k_0 \ ,$$

which contradicts our assumption that $n \geq 6$. ♢

CLAIM 8.7. L is a 1-bridge braid in V.

PROOF OF CLAIM 8.7. Recall that in Γ_P all the c_j's have the same sign. Claim 8.5 implies that all the b_i's have the same sign. Hence Γ_P has only CD-edges and Γ_Q has only BE-edges. Therefore there are five parallel BE-edges in Γ_Q; see Figure 8.2. Let $\gamma \subset \partial V$ be the arc in ∂Q shown in Figure 8.2. Then the bigon faces of Γ_Q shown in Figure 8.2 allow us to define a (non-ambient) isotopy of L in V taking it to $\gamma \cup \delta$, where δ is the image of $e \cup e_1'$ in $P \subset D$, where we treat the vertex c_1 as a point in D so that δ is considered as a properly embedded arc in D. See Figure 8.3 for a typical situation of the arcs e_1, e_2, e_3, e_4, e_1' in P. This shows that L is a 1-bridge braid in V (with δ as the bridge). (Note that since Y is hyperbolic L is not a 0-bridge braid.) ♢

We thus have a 1-bridge braid L in V with a Dehn filling (namely $\bar{\alpha}$-filling) on the boundary component ∂V of the exterior Y of L in V that gives a solid torus. Such braids are classified in [**Wu3**]. In particular [**Wu3**, Table 1] shows that there is a unique example (up to homeomorphism) where L has winding number 4 in V: L is the closure of the braid $\sigma_1(\sigma_3\sigma_2\sigma_1)^2$. This braid is conjugate to $\sigma_2\sigma_1\sigma_3\sigma_2\sigma_1^3$, whose closure is shown in Figure 8.4. It is clear from this figure that L bounds a Möbius band in V, contradicting the fact that Y is hyperbolic.

This contradiction completes the proof of Lemma 8.3. ♢

To finish the proof of Proposition 8.1 we only need to prove the following two lemmas.

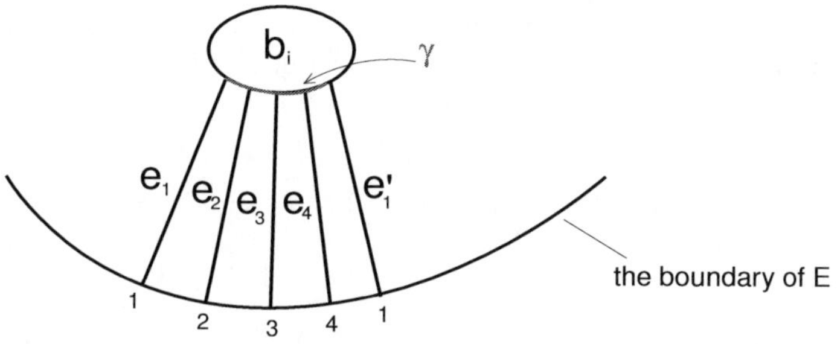

FIGURE 8.2. Five parallel adjacent BE-edges in Γ_Q

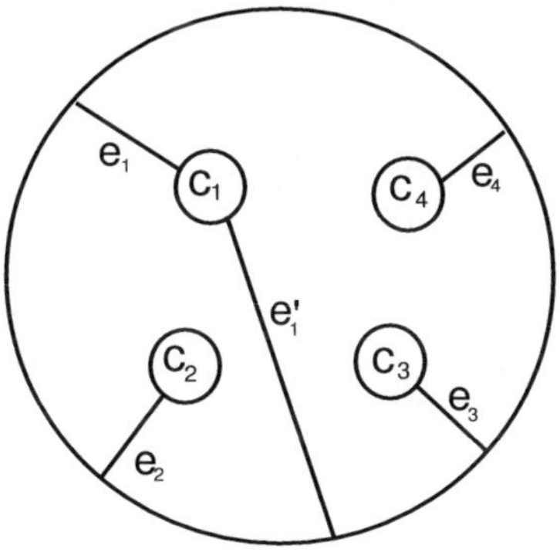

FIGURE 8.3. The five edges in Γ_P dual to the ones in Figure 8.2

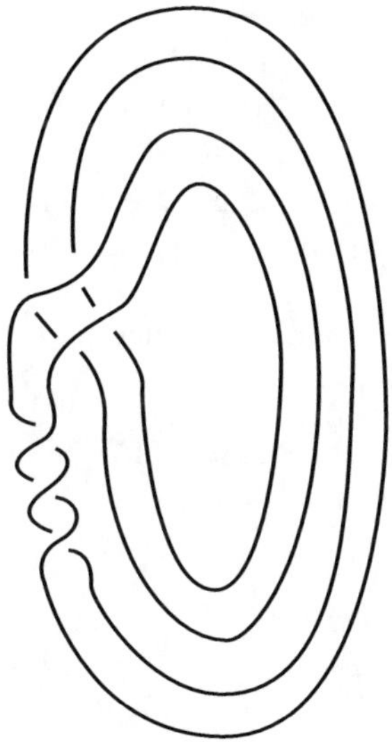

FIGURE 8.4. The unique 1-bridge 4-braid L

LEMMA 8.8. *Suppose that M is a hyperbolic knot manifold which contains an essential twice-punctured torus F of boundary slope β and let α be a slope on ∂M*

such that $M(\alpha)$ is an irreducible small Seifert manifold. If F is a fibre in M, $M(\alpha)$ is not a lens space, and τ_α preserves the orientations of the Seifert fibres of $M(\alpha)$, then $\Delta(\alpha, \beta) \leq 2$ when $M(\alpha)$ is not a prism manifold.

LEMMA 8.9. *Suppose that M is a hyperbolic knot manifold which contains an essential twice-punctured torus F of boundary slope β and let α be a slope on ∂M such that $M(\alpha)$ is a prism manifold. If F is a fibre in M and τ_α preserves the orientations of the Seifert fibres of $M(\alpha)$, then $\Delta(\alpha, \beta) \leq 3$.*

PROOF OF LEMMA 8.8. The argument is similar to that given in [**BGZ3**, §6].

Suppose that $\Delta = \Delta(\alpha, \beta) \geq 3$. Then the branch set L_α is a set of at most three Seifert fibres of $L(\bar{p}, \bar{q}) = M(\alpha)/\tau_\alpha$ with the induced Seifert fibration from $M(\alpha)$ given by [**BGZ3**, Lemma 4.3]. Since L is a hyperbolic link in V, K_α is not contained in L_α. Thus $L_\alpha = L$ so that q is odd and $L(\bar{p}, \bar{q})$ is a lens space of order $2p = 2\Delta \geq 6$. As in [**BGZ3**, §6.1], we

- define X to be the exterior of L in $V(\bar{\alpha}) = L(\bar{p}, \bar{q})$ and Y its exterior in V;
- fix a component K of L which is a regular fibre of $L(\bar{p}, \bar{q})$ and define T_K; respectively T_V, to be the component of ∂Y corresponding to K, respectively ∂V;
- use a meridian disk of V to construct a 4-punctured disk P properly embedded in Y;
- construct an essential, separating, vertical annulus $(A, \partial A) \subset (X, T_K)$ which separates X into two components X_1 and X_2 such that each X_i is Seifert with base orbifold either an annulus with no cone points or a disk with one cone point, of order at least 3 ([**BGZ3**, Lemma 6.1]).
- construct an n-punctured essential annulus Q in Y, where n is even, from an appropriately chosen essential separating vertical annulus in the exterior of K in $L(\bar{p}, \bar{q})$;
- define graphs Γ_P and Γ_Q, vertices d_V, $c_1, \ldots, c_4$, a_1, a_2, $b_1, \ldots, b_n$, and D-edges, CD-edges, etc.;

In the graph Γ_P, d_V has valency $2\Delta n \geq 6n$, and each c_j has valency 0 or 2.

The proof closely follows the proof of Case II of Chapter 7.

First assume $n \geq 4$. The argument immediately preceding Lemma 7.12 then shows that, with the notation established there,

$$\begin{aligned} 2\Delta n &\leq kn + 2(4 - k_0) \\ &\leq kn + 2(5 - k) \end{aligned}$$

giving

$$(2\Delta - k)n \leq 2(5 - k)\ .$$

Since $2\Delta \geq 6$, this is a contradiction.

So we may assume $n = 2$. Note that there are no parallel D-edges in Γ_P and valency $d_V \geq 12$. We now follow the proof of Lemma 7.19. By Lemma 7.7(1), $\ell = 2, 3$ or 4.

<u>$k = 5$</u>. The D-edges of Γ_P are as shown in Figure 8.5. Applying Lemma 7.14, the face f implies that c_3 and c_4 have valency 2, and similarly the face g implies that c_1 and c_2 have valency 2. Then d_V has valency 18, implying that $\Delta = 9$, a contradiction.

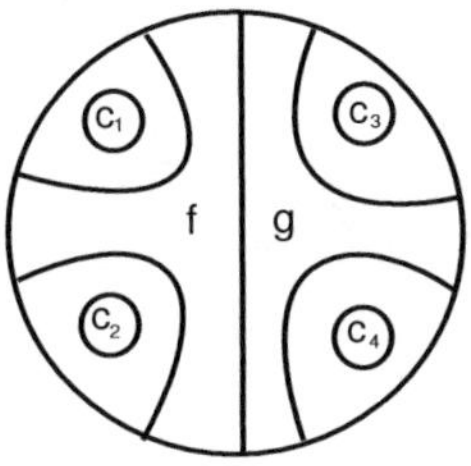

FIGURE 8.5. The D-edges of Γ_P when $k = 5$

$\underline{k=4}$. The D-edges of Γ_P are as shown in Figure 7.6(a) or (b).

In case (a) the proof of Lemma 7.19 applies.

In case (b) the proof of Lemma 7.19 applies unless c_1 and c_2 have valency 0. In this case, c_3 and c_4 have valency 2 and there are two possibilities for Γ_P, shown in Figure 8.6(i) and (ii). In both cases the faces f and g contradict Lemma 7.14.

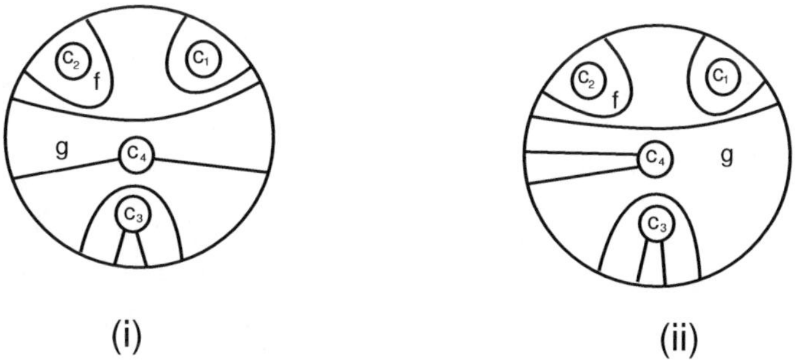

FIGURE 8.6. The two possibilities for Γ_P when c_1, c_2 have valency 0 and c_3, c_4 have valency 2 in Case (b) of $k = 4$.

$\underline{k=3}$. Since $12 \leq$ valency $d_V = 2k + 2\ell$ we have $\ell = 3$ or 4. The proof of Lemma 7.19 then applies to rule out every case here except that shown in Figure 7.12(iv). But in that case d_V has valency 10.

$\underline{k=2}$. Here $\ell = 4$, and the appropriate parts of the proof of Lemma 7.19 apply. More precisely, in case (a) take the part of the proof where c_3 and c_4 have valency 2. In case (b), take the part of the proof that starts with the supposition that both c_2 and c_3 have valency 2. Of the possibilities for Γ_P shown in Figure 7.16, (vi) and (vii) cannot occur here since valency $d_V \geq 12$. The other cases are eliminated as in the proof of Lemma 7.19.

$\underline{k=1}$. Since $12 \leq$ valency $d_V = 2k + 2\ell \leq 10$, this case is impossible. ◇

PROOF OF LEMMA 8.9. Suppose otherwise that $\Delta = \Delta(\alpha, \beta) \geq 4$. Then by [**L2**], $\Delta = 4$.

We may assume, by [**BGZ3**, Lemma 4.1], that the τ_α-invariant Seifert structure on $M(\alpha)$ is the one whose base orbifold is $S^2(2, 2, c)$ for some $c \geq 2$. We continue to use the notations established so far in this chapter. In the present case we still have the manifolds $X, T_K, T_V, A, X_1, X_2, Y, P, Q$ and the graphs Γ_P, Γ_Q defined as above in this chapter. The only new situation that possibly arises in the present case (which we assume happens since otherwise the argument of Lemma 8.8 works identically) is that exactly one of X_1 and X_2, say X_1, is a solid torus whose singular fibre has order two, and X_2 is either a $T^2 \times I$ or a solid torus whose singular fibre has order bigger than two. This assertion follows from the proof of [**BGZ3**, Lemma 6.1]. Thus the graph Γ_P may contain D-edge S-cycles, and the proof of Lemma 7.9 yields the following.

CLAIM 8.10. The bigon face bounded by a D-edge S-cycle in Γ_P lies on the X_1-side of A. $\diamondsuit$

CLAIM 8.11. When $n \geq 4$, Γ_P cannot have D-edge extended S-cycles.

PROOF OF CLAIM 8.11. Let $\{e_1, e_2, e_3, e_4\}$ be a D-edge extended S-cycle with label sequence $\{i-1, i, i+1, i+2\}$ (here labels are defined mod n). Let R_j be the bigon face between e_j, e_{j+1} for $j = 1, 2, 3$. Then R_2 is contained in X_1 by Claim 8.10, and R_1 and R_3 in X_2. Let H_j be the component of $V_{\bar{\alpha}} \setminus \cup_{k=1}^{n} \widehat{b}_k$ connecting $\{\widehat{b}_j, \widehat{b}_{j+1}\}$ for $j = i-1, i, i+1$. Let U_1 be a regular neighbourhood of $H_i \cup R_2$ in X_1, and U_2 a regular neighbourhood of $H_{i-1} \cup H_{i+1} \cup R_1 \cup R_3$ in X_2. Then U_1 is a solid torus and the frontier E_1 of U_1 in X_1 is an annulus with winding number 2 in U_1. Thus E_1 is parallel to $\partial X_1 \setminus (\partial U_1 \cap A)$ in $X_1 \setminus U_1$. The manifold U_2 is also a solid torus, the frontier of U_2 in X_2 is a pair of annuli E_2', E_2'', and E_2' is parallel to E_2'' in U_2. We may assume that $U_1 \cap A$ is equal to a component of $U_2 \cap A$. Let U_3 and U_4 be the two components of $X_2 \setminus U_2$, and we may assume that U_4 is the one which contains $\partial X_2 \setminus A$ and that $E_2'' \subset \partial U_4$ (cf. Figure 8.7). Note that U_4 must be a solid torus in which E_2'' is parallel to $\partial U_4 \setminus E_2''$ for otherwise the frontier of U_4 in X would be an essential annulus in X which has the same boundary as the annulus A in T_K but has less number of intersection components with $V_{\bar{\alpha}}$ than A. Thus U_3 is either a torus cross interval or a solid torus in which E_2' has winding number larger than two. Now let A' be the annulus $E_1 \cup E_2' \cup [A \setminus (U_1 \cup U_2)]$ (cf. Figure 8.7). Then A' is an essential annulus in X such that $\partial A' = \partial A$ in T_K but A' has less number of intersection components with $V_{\bar{\alpha}}$ than A, yielding a final contradiction. $\diamondsuit$

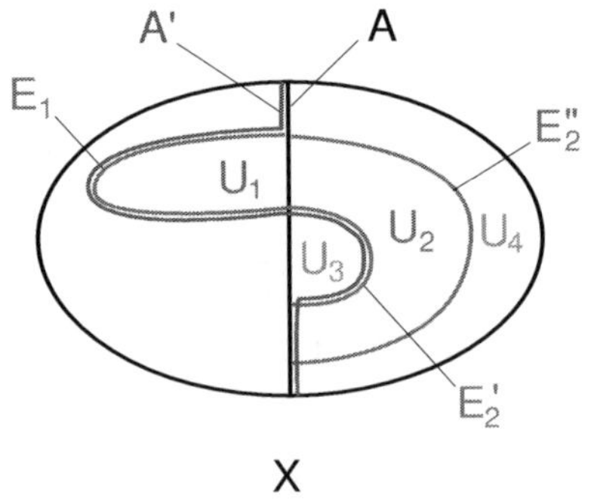

FIGURE 8.7. The submanifolds $A, A', U_1, U_2, U_3, U_4$ of X.

Note that Lemma 7.11 still holds in the present case.

CLAIM 8.12. Γ_P cannot have a family of more than $\frac{n}{2}+1$ parallel D-edges.

PROOF OF CLAIM 8.12. Otherwise Γ_P has either an D-edge extended S-cycle or two S-cycles with distinct label pairs, contradicting Claim 8.11 or Lemma 7.11 respectively. $\diamondsuit$

CLAIM 8.13. In Γ_Q, the endpoints of the AB-edges incident to b_i are consecutive around b_i, for each fixed $i = 1, \ldots, n$.

PROOF OF CLAIM 8.13. By the parity rule all the AB-edges incident to a fixed b_i have their other endpoint on a fixed component of $\partial A = \{a_1, a_2\}$. Also observe that all the b_i's which are connected to a fixed component of $\{a_1, a_2\}$ by AB-edges have the same sign, and thus there can be no B-edges connecting between these b_i's (because all B-edges are negative). Noticing that every vertex b_i is incident to some B-edges in Γ_Q, one can now see that the lemma follows. (cf. Lemma 7.17 and its proof.) $\diamondsuit$

CLAIM 8.14. The endpoints with label i on d_V of the CD-edges of Γ_P (when non-empty) are consecutive among all the edge-endpoints with label i on d_V, for each fixed $i = 1, \ldots, n$.

PROOF OF CLAIM 8.14. This follows from Claim 8.13 and the condition that $\Delta = 4$. (cf. Corollary 7.18 and its proof.) $\diamondsuit$

CLAIM 8.15. $n = 2$.

PROOF OF CLAIM 8.15. Since there are at most 8 CD-edges, there are at least $(8n-8)/2 = 4n-4$ D-edges in Γ_P. Thus there is a family of parallel D-edges in Γ_P with at least $(4n-4)/5$ edges. By Claim 8.12, $(4n-4)/5 \leq \frac{n}{2}+1$. Thus $n \leq 6$. So assume for contradiction that $n = 6$ or 4.

If $n = 6$, then there are exactly 8 CD-edges and 5 families of parallel D-edges, each having 4 edges. So part of Γ_P may be assumed to be as shown in Figure 8.8. So there is an extended D-edge S-cycle in Γ_P, contradicting Claim 8.11.

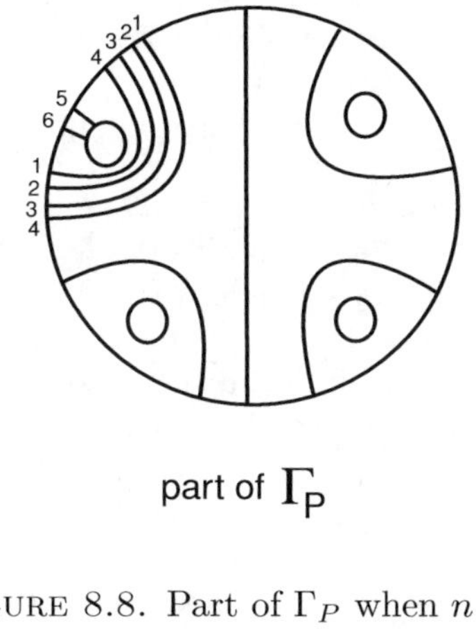

FIGURE 8.8. Part of Γ_P when $n = 6$

So $n = 4$. Note that the total weights of all D-edges in $\overline{\Gamma}_P$ is at least 12 and each D-edge of $\overline{\Gamma}_P$ has weight at most 3 by Claim 8.12. It follows that $\overline{\Gamma}_P$ has no C-edges. It also follows that there is an D-edge $\bar{e}$ in $\overline{\Gamma}_P$ of type O and with weight

3. Let c_j be the single vertex that $\bar{e}$ cuts off as given in the definition of an O-type edge. We note that there must be two CD-edges incident to c_j. For otherwise a part of Γ_P is as shown in Figure 8.9. In the figure the face f is bounded by an D-edge S-cycle and thus is contained in X_1 (which is a solid torus), but the face g is also contained in X_1, yielding a contradiction since it abuts a puncture c_i, which cannot exist in the solid torus X_1.

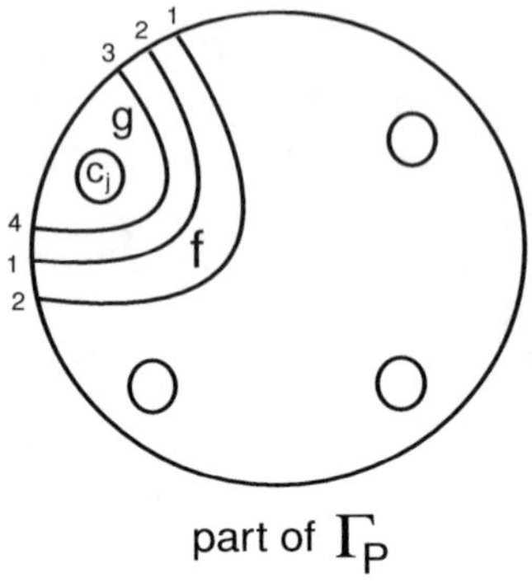

FIGURE 8.9. Part of Γ_P when $n = 4$ and c_j has valency 0.

Hence part of Γ_P is as shown in Figure 8.10(a). If there is no N-type D-edge in $\overline{\Gamma}_P$, then there are exactly 8 CD-edges and 12 D-edges in Γ_P, and Γ_P must look like that shown in Figure 8.10(b). But then we get a contradiction with Claim 8.14, considering the label 1. Therefore part of Γ_P maybe assumed as shown in Figure 8.10(c) where the edge e has label 4 at its upper endpoint.

By the parity rule the lower endpoint of the edge e has label 1 or 3. In the former case the part of the graph Γ_P on the left-hand side of the edge e must be as shown in Figure 8.10(d) or (e), by considering labels around d_V. But then the face g shown in both of the subcases is bounded by an D-edge Scharlemann cycle of order 3 and lies on X_1-side, contradicting Lemma 7.11. In the latter case the left-hand side of the edge e must be as shown in Figure 8.10(f). But then we get a contradiction with Claim 8.14 by looking at the label 4. $\diamondsuit$

So $n = 2$. By the proof of Lemma 8.8 we may only consider the situation when Γ_P contains D-edge S-cycles, i.e. Γ_P contains parallel D-edges. But note that each parallel family of D-edges contains at most two edges by Claim 8.10. We have at least 4 D-edges in Γ_P.

Note that Lemma 7.14 is still valid in the present case.

Recall that k is the number of D-edges in $\overline{\Gamma}_P$ and $k \leq 5$. In the current case $k \geq 2$. Note that we have assumed that there are at least $k+1$ D-edges in Γ_P.

If $k = 4$ or 5, then D-edges in $\overline{\Gamma}_P$ are as shown in Figure 7.5 or Figure 7.6, where at least one edge has weight 2. One can easily see from the figures that Γ_P must have, in each case, a disk face bounded by an D-edge Scharlemann cycle of order 3 or 4 that lies in X_2. Hence X_2 is a solid torus by Lemma 7.14 and thus there are 8 CD-edges in Γ_P. But there are at least 5 D-edges in Γ_P. So the valency of the vertex d_V would be at least 18 in Γ_P, contradicting to the assumption that the valency of d_V is 16 in Γ_P.

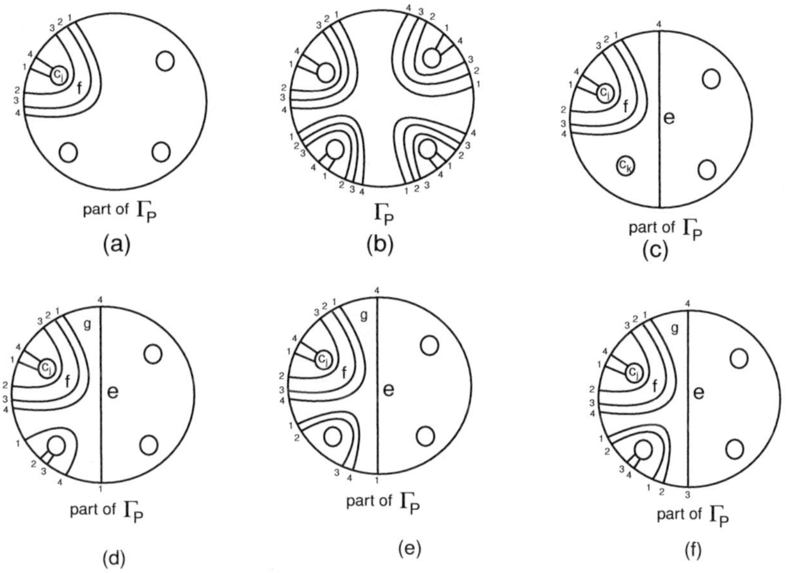

FIGURE 8.10. Six possibilities for Γ_P when $n = 4$ and c_j has valency 2.

If $k = 3$, then the three D-edges in $\overline{\Gamma}_P$ are as shown in Figure 7.9 (a) (b) (c), with at least one edge having weight 2. We first consider Case (a) of Figure 7.13. If the vertex c_4 has valency 2 in Γ_P, then Γ_P has a disk face with at least two d_V-corners, contained in X_2. Hence X_2 is a solid torus and every c_i has valency two in Γ_P. Now it is easy to see that Claim 8.14 is violated. So the valency of c_4 is zero, and thus there are at least 5 D-edges in Γ_P. Since there are at most 6 D-edges in Γ_P, at least two of c_1, c_2, c_3 have valency 2. Now it is easy to see that we have a contradiction again with Claim 8.14. Next we consider Case (b) of Figure 7.13. In this case Γ_P has a disk face bounded by an D-edge Scharlemann cycle of order 3, contained in X_2. Hence X_2 is a solid torus by Lemma 7.14 and thus every c_i has valency 2 in Γ_P. Now one can see a contradiction with Claim 8.14. Lastly in Case (c) of Figure 7.13 one can easily see a contradiction with Claim 8.14 once again.

If $k = 2$, then the two D-edges in $\overline{\Gamma}_P$ are as shown in Figure 7.13 (a) (b), each having weight 2. Also there are exactly 8 CD-edges in Γ_P. Now it is easy to see that Claim 8.14 is violated in both (a) and (b) cases. ◇

CHAPTER 9

Further assumptions, reductions, and background material

The remainder of the paper deals with the cases that F is neither a fibre or semi-fibre.

LEMMA 9.1. *If F is neither a fibre nor semi-fibre in M and $b_1(M) \geq 2$, then*

$$\Delta(\alpha, \beta) \leq \begin{cases} 1 & \text{if } M(\alpha) \text{ is very small} \\ 5 & \text{otherwise} \end{cases}$$

PROOF. If $b_1(M) \geq 2$, there is a closed, non-separating, Thurston norm minimizing surface S in the interior of M. By [**Ga**, Corollary, page 462], S remains Thurston norm minimizing in $M(\gamma)$ for all but at most one slope γ_0 on ∂M. If $\alpha = \gamma_0$, a result of Wu implies that $\Delta(\alpha, \beta) \leq 1$ (cf. [**BGZ1**, Proposition 5.1]), so suppose that $\alpha \neq \gamma_0$. Then $M(\alpha)$ cannot be very small since it contains S as an essential surface. Theorem 1.2 of [**BGZ1**] now shows that $\Delta(\alpha, \beta) \leq 5$. ♢

We will apply the results of [**BGZ2**] in what follows, which requires that more care be taken in the choice of F. Here is a list of assumptions we will make in our proof of Theorem 1.2 when F is neither a fibre nor semi-fibre. The references which justifiy the sufficiency of these assumptions are listed as well.

ASSUMPTIONS 9.2.

(1) $b_1(M) = 1$ (cf. Lemma 9.1).
(2) M does not admit a fibre or semi-fibre which is an essential twice-punctured torus of boundary slope β (cf. Propositions 7.1 and 8.1).
(3) M admits no essential once-punctured torus of boundary slope β (cf. [**BGZ3**, Theorem 1.3]).
(4) F is chosen according to the assumptions of [**BGZ2**, §2].
(5) If M_F is not connected, then it is a union of two genus 2 handlebodies (cf. [**BGZ3**, Proposition 3.3]).

Next we provide a summary of the notation and terminology from [**BGZ2**] that will be used below.

Let F be chosen as above. If F separates M we take S to be F. Otherwise we take S to be the frontier of a small radius tubular neighbourhood of F in M. Thus S consists of two parallel copies F_1, F_2 of F. In either case S splits M into two components X^+ and X^-. By Assumption 9.2 (2) above, we can suppose that X^+ is not an I-bundle.

Let $\widehat{S}$ be a closed surface in $M(\beta)$ obtained by attaching disjoint meridian disks of the β-filling solid torus to S. Then $\widehat{S}$ splits $M(\beta)$ into two compact submanifolds $\widehat{X}^+$ containing X^+ and $\widehat{X}^-$ containing X^-, each having incompressible boundary $\widehat{S}$.

It was shown in [**BCSZ1**] how to construct an immersion $h : Y \to M(\alpha)$ where Y is a disk or torus, a labeled "intersection" graph $\Gamma_F = h^{-1}(F) \subset Y$, and, for each sign $\epsilon = \pm$, a sequence of characteristic subsurfaces

$$F = \dot{\Phi}_0^\epsilon \supseteq \dot{\Phi}_1^\epsilon \supseteq \dot{\Phi}_2^\epsilon \supseteq \ldots \supseteq \dot{\Phi}_n^\epsilon \supseteq \ldots$$

See [**BCSZ1**, §5] for the definition of the j-th characteristic subsurface $\Phi_j^\epsilon \subseteq S$. We shall assume throughout the paper that Φ_j^ϵ is neatly embedded in S ([**BGZ2**, §3.1]). It is characterized up to ambient isotopy by the following property:

$$(*)\begin{cases} \text{a large function } f_0 : K \to S \text{ admits an essential homotopy of length } j, \\ \text{which starts on the } \epsilon\text{-side of } S \text{ if and only if it is homotopic in } S \text{ to a map} \\ \text{with image in } \Phi_j^\epsilon \end{cases}$$

See [**BCSZ1**, Proposition 5.2.8]. When $j = 1$, basic Jaco-Shalen-Johannson theory guarantees the existence of an (I, S^0)-bundle pair $(\Sigma_1^\epsilon, \Phi_1^\epsilon) \subset (X^\epsilon, S)$. It was shown in [**BGZ2**, Proposition 4.9] that we can assume that $(\Sigma_1^\epsilon, \Phi_1^\epsilon)$ is neatly embedded in (X^ϵ, S) ([**BGZ2**, §3.2]).

As in [**BGZ2**, §3.2] we take $\dot{\Phi}_j^\epsilon$ to be the union of the components of Φ_j^ϵ which contain some outer boundary components and $\breve{\Phi}_j^\epsilon$ to be the neat subsurface in S obtained from the union of $\dot{\Phi}_j^\epsilon$ and a closed collar neighbourhood of $\partial S - \partial\dot{\Phi}_j^\epsilon$ in $S - \dot{\Phi}_j^\epsilon$. There are corresponding I-bundle pairs $(\dot{\Sigma}_1^\epsilon, \dot{\Phi}_1^\epsilon)$ and $(\breve{\Sigma}_1^\epsilon, \breve{\Phi}_1^\epsilon)$ neatly embedded in (X^ϵ, S).

A neat subsurface S_0 of S is called *tight* if it caps off to a disk in $\widehat{S}$. Equivalently, S_0 is a connected, planar, neat subsurface of S with one inner boundary component.

We use t_j^ϵ to denote the number of tight components of $\breve{\Phi}_j^\epsilon$. If j is odd, t_j^ϵ is even, while if j is even, $t_j^+ = t_j^-$. See [**BGZ2**, §6].

An *intersection graph* Γ_F can be constructed in a disk or torus Y from an immersion

$$h : Y \to M(\alpha) \tag{9.1}$$

(cf. [**BGZ2**, Section 11]). The immersion maps the vertices of Γ_F to meridian disks of the α-filling torus, edges of Γ_F to F, and faces of Γ_F to X^+ or X^-. For simplicity we shall say a face of Γ_F is contained in X^ϵ if its image under the immersion is contained there. We refer to [**BGZ2**, Sections 11 and 12] for terms, notations and basic facts concerning Γ_F.

REMARK 9.3. Note that the intersection graph Γ_F in Y has the following property: either Γ_F has a connected component which lies in a subdisk of Y or Y is a torus and every face of Γ_F is a disk or an annulus. So we may and shall assume that either Γ_F is a connected graph in a disk or it is a graph in a torus with only disk faces and/or annulus faces. It turns out that our graph related arguments are never affected whether annulus faces exist or not.

CHAPTER 10

The proof of Theorem 1.2 when F is non-separating but not a fibre

In this chapter we prove that part of Theorem 1.2 dealing with the case that F is non-separating but not a fibre. We suppose throughout that Assumptions 4.3 and 9.2 hold. Note that we can assume that the components of ∂F are like-oriented on ∂M. For otherwise we can attach a peripheral annulus to F to obtain a closed non-separating genus two surface in M. This surface must be incompressible in M since M is hyperbolic. It follows from [**CGLS**, Theorem 2.4.3] that β is a singular slope, which contradicts Assumptions 4.3(4). By construction, $X^- = F \times I$.

We shall prove Proposition 10.1.

PROPOSITION 10.1. *Suppose that M is a hyperbolic knot manifold which contains an essential, non-separating twice-punctured torus F of boundary slope β and let α be a slope on ∂M such that $M(\alpha)$ is an irreducible small Seifert manifold. If F is not a fibre and Assumptions 4.3 and 9.2 hold, then*

$$\Delta(\alpha,\beta) \leq \begin{cases} 4 & \text{if } M(\alpha) \text{ is very small} \\ 5 & \text{otherwise} \end{cases}$$

We begin with a result whose proof follows from [**BGZ2**].

PROPOSITION 10.2. *Suppose that F is a non-separating, essential, twice-punctured torus of slope β in a hyperbolic knot manifold M which completes to an essential torus in $M(\beta)$ but is not a fibre in M. Suppose as well that $M(\alpha)$ is an irreducible small Seifert manifold.*

(1) *If $t_1^+ > 0$, then* $\Delta(\alpha,\beta) \leq \begin{cases} 3 & \text{if } M(\alpha) \text{ is very small} \\ 4 & \text{otherwise} \end{cases}$

(2) *If $t_1^+ = 0$, then*

(a) $\Delta(\alpha,\beta) \leq \begin{cases} 4 & \text{if } M(\alpha) \text{ is very small} \\ 6 & \text{otherwise} \end{cases}$

(b) *$M(\beta)_{\widehat{F}}$ admits a Seifert structure with base orbifold an annulus with one cone point.*

PROOF. Assertion (2)(b) of the proposition is a consequence of [**BGZ2**, Lemma 7.10]. Assertion (2)(a) follows from [**BGZ2**, Propositions 13.1 and 13.2], as does the general inequality $\Delta(\alpha,\beta) \leq 4$ when $t_1^+ > 0$ claimed in assertion (1).

Assume that $t_1^+ > 0$ and $M(\alpha)$ is very small. Since t_1^+ is even and the number of boundary components F is bounded below by $\frac{t_1^+}{2}$, we have $t_1^+ \in \{2,4\}$. As $M(\alpha)$ is very small, the graph $\overline{\Gamma}_S$ is contained in a 2-disk and so it has a vertex of valency 5 or less (e.g. see [**BGZ2**, Proposition 12.2 and Corollary 12.4]). Hence if $t_1^+ = 4$, then [**BGZ2**, Inequality 13.0.1] shows that $\Delta(\alpha,\beta) \leq 2$. Suppose then that $t_1^+ = 2$ and note that $\overline{\Gamma}_S$ has a vertex of valency 3 or less, for if it doesn't, [**BGZ2**, Lemma

11.6 and Proposition 11.5(2)] imply that $\overline{\Gamma}_S$ is rectangular and so is contained in a torus, contrary to our assumptions. But then [**BGZ2**, Inequality 13.0.1] shows that $\Delta(\alpha, \beta) \leq 3$, so we are done. $\diamondsuit$

10.1. Proof of Proposition 10.1 when $M(\alpha)$ is not very small

By Proposition 10.2 we can suppose that $t_1^+ = 0$ in this section and the base orbifold $S^2(a, b, c)$ of $M(\alpha)$ is hyperbolic.

Recall from Chapter 9 that S is the frontier of X^+ in M and consists of two parallel copies F_1, F_2 of F. By Proposition 10.2 we can assume that $\Delta(\alpha, \beta) = 6$ in order to obtain a contradiction. In this case, the proof of [**BGZ2**, Proposition 13.2] shows that the reduced graph $\overline{\Gamma}_S$ is rectangular with every edge having weight 6, both $\dot{\Phi}_3^+$ and $\dot{\Phi}_5^-$ consist of a pair of tight components, each a twice-punctured disk, $\dot{\Phi}_5^+$ is a collar on ∂S, and so contains no large components.

Let $b_1, \ldots, b_4$ denote the components of $\partial S = \partial F_1 \cup \partial F_2$ indexed as they appear successively along ∂M and where $b_1 \cup b_3 = \partial F_1$ and $b_2 \cup b_4 = \partial F_2$. These four circles cut ∂M into four annuli $A_{i,i+1}, i = 1, \ldots, 4$, such that $\partial A_{i,i+1} = b_i \cup b_{i+1}$ (indexed (mod 4)). We assume that $\partial X^+ = S \cup A_{2,3} \cup A_{4,1}$.

As in [**BGZ3**], an *n-gon* in X^+ means a singular disk D with $\partial D \subseteq \partial X^+$ such that $\partial D \cap (A_{2,3} \cup A_{4,1})$ is a set of n embedded essential arcs in $A_{2,3} \cup A_{4,1}$, called the *corners* of D, and $\partial D \cap S$ is a set of n singular arcs, called the *edges* of D. As we go around ∂D in some direction we get a cyclic sequence of $X_2^{\pm 1}$ and $X_4^{\pm 1}$-*corners*, where X_2, X_2^{-1} indicate that ∂D is running across $A_{2,3}$ from 2 to 3 or from 3 to 2, respectively, and X_4, X_4^{-1} indicate that ∂D is running across $A_{4,1}$ from 4 to 1 or 1 to 4, respectively. In this way D determines a cyclic word $W = W(X_2^{\pm 1}, X_4^{\pm 1})$, well-defined up to inversion, and we say that D is *of type* W. (Thus D is of type W if and only if it is of type W^{-1}.) We emphasize that W is an unreduced word; for example X_2 and $X_2 X_4 X_4^{-1}$ are distinct.

There are no n-gons in X^+ with n odd (cf. [**BGZ3**, Lemma 11.6]).

LEMMA 10.3. *There is no bigon D in X^+ whose edges e_1, e_2 are essential paths in $(\dot{\Phi}_5^-, \partial S)$ and for which the inclusion $(D, e_1 \cup e_2) \to (X^+, \dot{\Phi}_5^-)$ is essential as a map of pairs.*

PROOF. Suppose otherwise that such a bigon D exists. Then D gives rise to an essential homotopy between its two edges and thus the edges of D can be homotoped, relative to their end points, into $\dot{\Phi}_1^+$. Then the essential intersection $\dot{\Phi}_5^- \wedge \dot{\Phi}_1^+$ contains a large component and therefore so does $\dot{\Phi}_6^+ = \tau_+(\dot{\Phi}_5^- \wedge \dot{\Phi}_1^+)$, contrary to the fact that $\dot{\Phi}_5^+$ has no large components. $\diamondsuit$

Recall that h is the π_1-injective map from the torus T into $M(\alpha)$ which induces the graph Γ_S in T. For a subset s of T we use s^* to denote its image under the map h. The image under h of every edge of a rectangular face of Γ_S is contained in $\dot{\Phi}_5^-$.

For notational simplicity, let us write $\dot{\Phi}_5^- = Q$, a pair of twice-punctured disks. Obviously we have $b_1 \cup b_3$ is contained in one component of Q and $b_2 \cup b_4$ in the other.

A singular disk $D \subset X^+$ whose edges are contained in Q will be called a *Q-disk*. An *essential* Q-disk is a Q-disk D such that each edge of D is an essential arc in Q.

Note that a Q-disk has no 12-, or 14-, or 32- or 34-edge. This simple fact will be used many times in the rest of the proof in determining the type $W(D)$ of a Q-disk D.

LEMMA 10.4. *An essential Q-n-gon, $n \leq 4$, is a 4-gon of type $X_2X_4^{-1}X_4X_4^{-1}$ or $X_4X_2^{-1}X_2X_2^{-1}$.*

PROOF. Let E be an essential Q-n-gon, $n \leq 4$. By the loop theorem ([**He**, Theorem 4.10]) we get an essential embedded Q-disk D, with $\{\text{corners of } D\} \subset \{\text{corners of } E\}$. Thus D is a k-gon with $k \leq 4$. We know k must be even, and D cannot be a bigon by Lemma 10.3. Hence D is a 4-gon.

There are three possibilities: D has either

(A) *all X_2-corners (or all X_4-corners);*
(B) *two X_2-corners and two X_4-corners;*
(C) *one X_2-corner and three X_4-corners (or vice versa).*

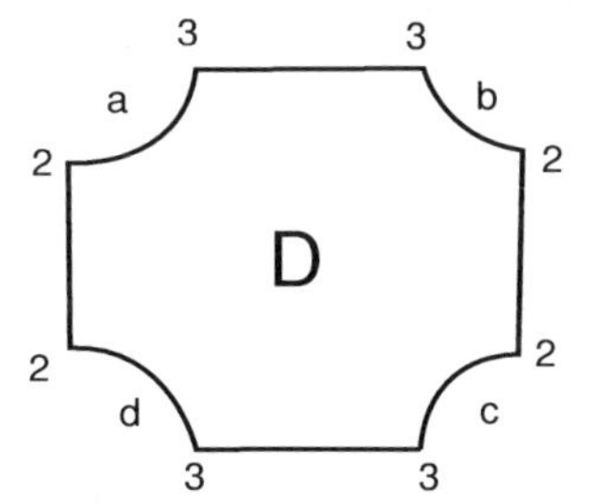

FIGURE 10.1. The labeled corners of the 4-gon D

In Case (A), $W(D) = X_2X_2^{-1}X_2X_2^{-1}$ (or $X_4X_4^{-1}X_4X_4^{-1}$). Label the corners of D a, b, c, d as shown in Figure 10.1. Then ∂D is as shown in Figure 10.2. Let $Y = \widehat{X}^- \cup H_{(23)}$. Note that ∂Y is a surface of genus 2. We see from Figure 10.2 that ∂D is isotopic in ∂Y to a meridian of $H_{(23)}$, and so bounds a non-separating disk $D' \subset Y$. Then $D \cup D'$ is a non-separating 2-sphere in $M(\beta)$, a contradiction.

In Case (B), the only possibilities for $W(D)$ are $X_2X_2^{-1}X_4X_4^{-1}$ and $X_2X_4^{-1}X_2X_4^{-1}$. In the first case, ∂D also contains a 1-loop and a 3-loop, which must intersect. In the second case, let $U = \widehat{Q} \times I \cup H_{(23)} \cup H_{(41)} \cup N(D) \subset \widehat{X}^+$. Then U is a punctured projective space in X^+, a contradiction.

Hence Case (C) must hold; so suppose that D has one X_2-corner and three X_4-corners. Since $\{\text{corners of } D\} \subset \{\text{corners of } E\}$, E is also a 4-gon with one X_2-corner and three X_4 corners. The possibility for $W(E)$ is $X_2X_4^{-1}X_4X_4^{-1}$. This completes the proof of Lemma 10.4 ♢

LEMMA 10.5. *There do not exist disjoint Q-disks of types $X_2X_4^{-1}X_4X_4^{-1}$ and $X_4X_2^{-1}X_2X_2^{-1}$.*

PROOF. Let D_1, D_2 be Q-disks of types $X_2X_4^{-1}X_4X_4^{-1}$ and $X_4X_2^{-1}X_2X_2^{-1}$, respectively. Then ∂D_1 contains a 1-loop and ∂D_2 contains a 3-loop, and these must intersect. ♢

LEMMA 10.6. *There cannot be essential Q-4-gons of both types $X_2X_4^{-1}X_4X_4^{-1}$ and $X_4X_2^{-1}X_2X_2^{-1}$.*

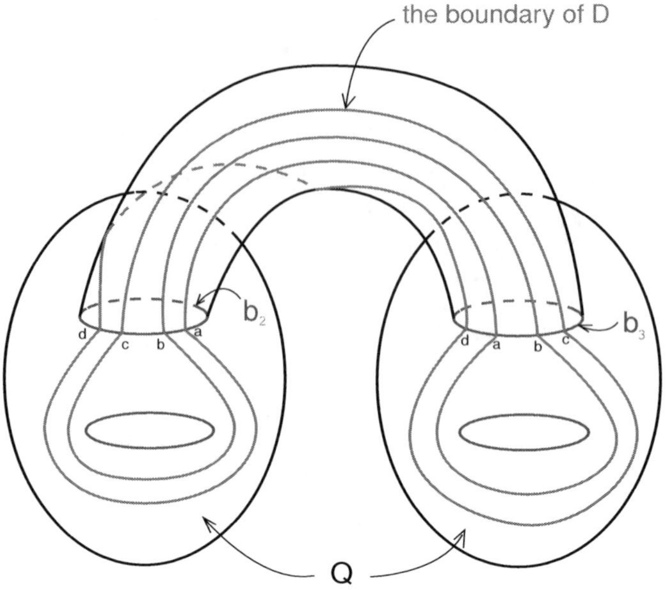

FIGURE 10.2. The boundary of D in Case (A)

PROOF. Let E_1, E_2 be Q-disks of types $X_2X_4^{-1}X_4X_4^{-1}$ and $X_4X_2^{-1}X_2X_2^{-1}$ respectively. By the loop theorem and Lemma 10.4 we get embedded Q-disks D_1 and D_2 of these types. By Lemma 10.5, D_1 and D_2 must intersect; consider an arc of intersection, coming from the identification of arcs $u_i \subset D_i$, $i = 1, 2$. We may assume that the endpoints of u_i lie on distinct edges of D_i, $i = 1, 2$. Then u_i separates D_i into two disks, D_i' and D_i'', say, where D_i' contains either one or two corners of D_i.

If D_1' and D_2' each contain a single corner, and these corners are distinct, then $D_1' \cup D_2'$ is a Q-bigon with one X_2- and one X_4-corner, contradicting Lemma 10.3.

If D_1' and D_2' both contain, say, a single X_2-corner, then u_1 is as shown in Figure 10.3, which also shows one of the three possibilities for u_2. Since b_1 and b_3 lie in one component of Q, say Q_1, and b_2 and b_4 lie in the other component, say Q_2, and each of the arcs u_1 and u_2 has one endpoint in Q_1 and one in Q_2, u_1 and u_2 must be identified as shown in Figure 10.3. Then $D_1^* = D_1'' \cup D_2'$ is a Q-disk of type $X_2X_4^{-1}X_4X_4^{-1}$ having fewer intersections than D_1 with D_2.

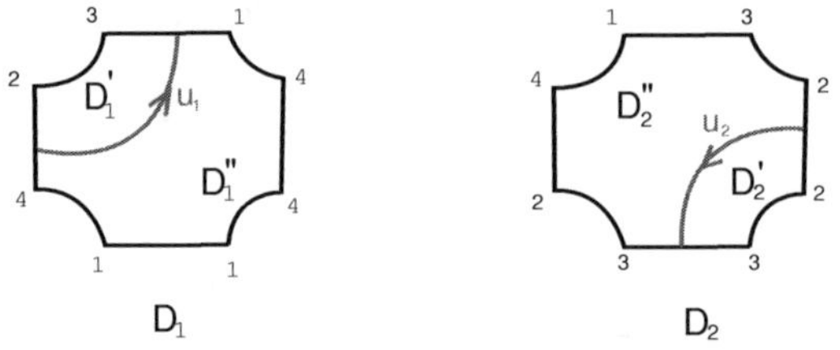

FIGURE 10.3. u_1, u_2 when both $D_{1'}, D_{2'}$ contain a single X_2-corner.

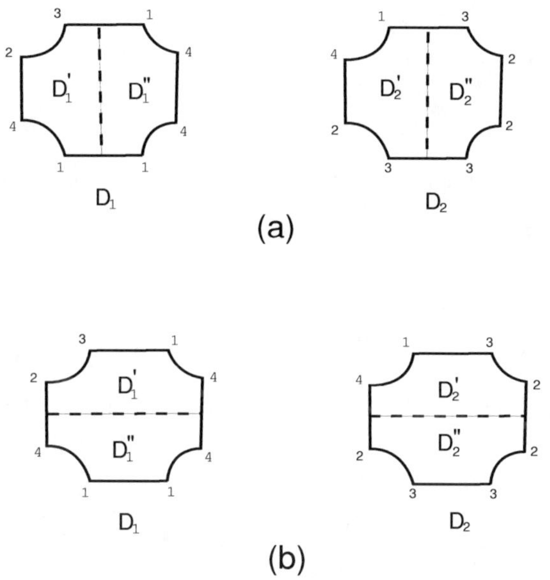

FIGURE 10.4. Possibilities when each of $D_{1'}, D_{1''}, D_{2'}, D_{2''}$ contains two corners.

If each of D_i' and D_i'' contains two corners, $i = 1, 2$, the two possibilities for u_1 and u_2 are illustrated in Figure 10.4, (a) and (b). In both cases, $D_1^* = D_1'' \cup D_2'$ is again a Q-disk of type $X_2X_4^{-1}X_4X_4^{-1}$ having fewer intersections with D_2.

Applying the loop theorem to the disk D_1^* constructed above, and using Lemma 10.4, we get an embedded Q-disk of type $X_2X_4^{-1}X_4X_4^{-1}$ having fewer intersections with D_2 than D_1. Continuing, we eventually get disjoint embedded Q-disk of types $X_2X_4^{-1}X_4X_4^{-1}$ and $X_4X_2^{-1}X_2X_2^{-1}$, contradicting Lemma 10.5. This completes the proof of Lemma 10.6. $\diamondsuit$

Since each edge of $\bar{\Gamma}_S$ has weight 6, consecutive 4-gon corners of Γ_S at a given vertex are distinct. Hence the total number of X_2-corners in the 4-gon faces of Γ_S is the same as the total number of X_4-corners. Since a 4-gon face of Γ_S is an essential Q-disk, this contradicts Lemmas 10.4 and 10.6.

10.2. Refinements to the very small case when $t_1^+ = 0$

In this section we refine the very small case of Proposition 10.1 when $t_1^+ = 0$. Assume that this is the case and recall that $M(\beta)_{\widehat{F}}$ admits a Seifert structure with base orbifold an annulus with one cone point of order $n \geq 2$ (cf. Proposition 10.2). Let T_0 and T_1 be the boundary components of $M(\beta)_{\widehat{F}}$ and ϕ_0, ϕ_1 the Seifert slope of $M(\beta)_{\widehat{F}}$ on T_0, T_1 respectively. Denote by $f : T_0 \to T_1$ the gluing map which produces $M(\beta)$ and set

$$d = \Delta(f_*(\phi_0), \phi_1)$$

PROPOSITION 10.7. *Let M be a hyperbolic knot manifold which contains an essential non-separating twice-punctured torus F of boundary slope β and let α be a slope on ∂M such that $M(\alpha)$ is a very small Seifert manifold. If $t_1^+ = 0$ and $d \neq 1$,* *then*

$$\Delta(\alpha,\beta) \le \begin{cases} 1 & \text{if } d=0 \text{ or } \alpha \text{ is of } C\text{-type} \\ 2 & \text{if } \alpha \text{ is of } D, T \text{ or } O\text{-type, or } (a,b,c) \text{ is } (2,3,6) \text{ or } (3,3,3) \\ 3 & \text{if } \alpha \text{ is of } I\text{-type or } (a,b,c) = (2,4,4) \end{cases}$$

PROOF. If $d = 0$, $M(\beta)$ is a Seifert fibred manifold with base orbifold a torus or Klein bottle with one cone point and therefore β is a singular slope of a closed essential surface in M (cf. the third paragraph of §4). Proposition 4.1 then shows that $\Delta(\alpha,\beta) \le 1$.

Assume that $d \ge 2$. By construction,

$$\pi_1(M(\beta)) \cong \langle \pi_1(M(\beta)_{\widehat{F}}), t : t\gamma t^{-1} = f_*(\gamma) \text{ for all } \gamma \in \pi_1(T_0)\rangle$$

Fix bases $\{\phi_j, \phi_j^*\}$ of $\pi_1(T_j)$ $(j = 0,1)$. Then the quotient of $\pi_1(M(\beta)_{\widehat{F}})$ by the fibre class is isomorphic to $\mathbb{Z} * \mathbb{Z}/n$ where there are generators a, b of $\mathbb{Z}, \mathbb{Z}/n$ respectively so that ϕ_0^* is sent to a and ϕ_1^* to ab.

Write $f_* = \begin{pmatrix} p & q \\ r & s \end{pmatrix}$ with respect to the bases $\{\phi_j, \phi_j^*\}$. Since $d = \Delta(f_*(\phi_0), \phi_1)$ $= |r|$, the integers d and s are coprime.

Since $d = \Delta(f_*(\phi_0), \phi_1)$, the quotient of $\pi_1(\widehat{F})$ by $\langle \phi_0, \phi_1\rangle$ is isomorphic to $\mathbb{Z}/d$ generated by either ϕ_0^* or ϕ_1^*. Hence if we quotient $\pi_1(M(\beta))$ by the normal closure of $\langle \phi_0, \phi_1\rangle$ in $\pi_1(M(\beta))$ we obtain an epimorphism

$$\varphi : \pi_1(M(\beta)) \to \langle a, b, t : a^d = 1, b^n = 1, (ab)^d = 1, tat^{-1} = (ab)^s\rangle = \Delta(d,n,d)*_\psi$$

where $\psi : \langle a\rangle \xrightarrow{\cong} \langle ab\rangle, a \mapsto (ab)^s$ (cf. §6.3). If $\beta^* \in \pi_1(\partial M)$ is a dual class to β, then

$$\varphi(\beta^*) = ta^j ta^l$$

for some integers j, l.

Let $Y_0 \subset X_{PSL_2}(\Delta(d,n,d)*_\psi)$ be a curve constructed in Lemma 6.4(3) by bending a representation (θ, A) and define

$$X_0 = (\varphi)^*(Y_0) \subset X_{PSL_2}(M(\beta)) \subset X_{PSL_2}(M)$$

Part (3) of Lemma 6.4 then shows that:

- X_0 is strictly non-trivial;
- $\tilde{f}_{\beta^*}$ has a pole at each ideal point of X_0;
- if either $n \ne 2$ or $d \notin \{2,4\}$, then X_0 has exactly two ideal points, and therefore $s_{X_0} \ge 2$;
- if $n = 2$ and $d \in \{2,4\}$, then X_0 has exactly one ideal point, and therefore $s_{X_0} \ge 1$.

By (5.2) we have

$$\|\alpha\|_{X_0} = \Delta(\alpha,\beta)s_{X_0}$$

If $\pi_1(M(\alpha))$ is cyclic, then [**BCSZ2**, Proposition 8.1] implies that $\Delta(\alpha,\beta)s_{X_0} = \|\alpha\|_{X_0} = s_{X_0}$ and therefore $\Delta(\alpha,\beta) = 1$. It also implies that $\Delta(\alpha,\beta)s_{X_0} = \|\alpha\|_X \le 2s_{X_0}$ if $M(\alpha)$ is a prism manifold since X_0 contains strictly irreducible characters. Thus $\Delta(\alpha,\beta) \le 2$ (cf. [**BCSZ2**, Proposition 8.1]).

To deal with the remaining cases, note that by Lemma 5.3 of [**BZ1**] and Propositions 5.2, 5.3 and 5.4 of [**Bo**], if $\rho : \pi_1(M(\alpha)) \to PSL_2(\mathbb{C})$ is an irreducible representation, then

$$\text{image}(\rho) \cong \begin{cases} T_{12} & \text{if } \alpha \text{ is of } T\text{-type} \\ D_3 \text{ or } O_{24} & \text{if } \alpha \text{ is of } O\text{-type} \\ I_{60} = \Delta(2,3,5) & \text{if } \alpha \text{ is of } I\text{-type} \\ D_3 \text{ or } T_{12} & \text{if } (a,b,c) = (2,3,6) \\ D_2 \text{ or } D_4 & \text{if } (a,b,c) = (2,4,4) \\ T_{12} & \text{if } (a,b,c) = (3,3,3) \end{cases}$$

Further, in each of these six possibilities at most two such representations have isomorphic images and if two, either α has I-type and the image is I_{60} or $(a,b,c) = (2,4,4)$ and the image is D_4. Note, in particular, that the image of ρ is a finite group whose non-trivial elements have order $2, 3, 4$, or 5.

Assume that $\Delta(\alpha, \beta) > 1$. Then as in the proof of Proposition 7.3 we see that $J_{X_0}(\alpha) \neq \emptyset$ and is contained in X_0^ν. Further, the image χ_ρ of an element of $J_{X_0}(\alpha)$ is a smooth point of $X_{PSL_2}(M)$. We know, moreover, that $\rho(\alpha) = \pm I$, and so ρ induces irreducible homomorphism

$$\bar{\rho} : \pi_1(M(\alpha)) \to PSL(2, \mathbb{C})$$

Lemma 6.4(2) then shows that $n, d \in \{2,3,4,5\}$.

Suppose that either $d \notin \{2,4\}$ or $d = 4$ and $n > 2$. Then Lemma 6.4(2) shows that $\theta(\Delta(d,n,d))$ is generated by two elements of order larger than 2 and so is not conjugate into $\mathcal{N}$. It follows that X_0 contains no dihedral characters and by Lemma 5.3 of [**BZ1**] and Propositions 5.2, 5.3 and 5.4 of [**Bo**] we have $|J_{X_0}(\alpha)| \leq 2$ with equality only if α is of I-type. Since $s_{X_0} \geq 2$, Proposition 5.2(2) implies that

$$\Delta(\alpha, \beta) \leq 1 + \frac{2|J_{X_0}(\alpha)|}{s_{X_0}} \leq 3$$

with equality implying that α is of I-type.

Next suppose that $n = 2$ and $d \in \{2,4\}$. That is, $(d,n,d) = (2,2,2)$ or $(4,2,4)$. In either case the curve X_0 is constructed by bending a representation (θ, A) where $\theta : \Delta(d,n,d) \to \mathcal{K} \cong \Delta(2,2,2)$ and $A\theta(a)A^{-1} = \theta(ab)$. It is shown in the last paragraph of the proof of Lemma 6.4 that we can suppose that $\theta(a) = \pm \left(\begin{smallmatrix} i & 0 \\ 0 & -i \end{smallmatrix}\right)$ and $\theta(ab) = \pm \left(\begin{smallmatrix} 0 & 1 \\ -1 & 0 \end{smallmatrix}\right)$. Further, X_0 is parametrised by the characters of the representations (θ, A_x) where

$$A_x = \pm \begin{pmatrix} x & i/2x \\ ix & 1/2x \end{pmatrix}$$

Suppose that $\mathcal{K}$ is normal in the image of (θ, A_x). That is, this image is either D_2, D_4, T_{12}, or O_{24}. Since $A_x\theta(a)A_x^{-1} = \theta(ab)$, A_x must commute with $\theta(b) = \pm \left(\begin{smallmatrix} 0 & i \\ i & 0 \end{smallmatrix}\right)$. The reader will verify that this occurs if and only if $x = \pm\frac{1}{\sqrt{2}}$ or $x = \pm\frac{i}{\sqrt{2}}$ and if it does, the image of (θ, A_x) is isomorphic to D_4. This rules out the possibility that the image of (θ, A_x) is D_2, T_{12}, or O_{24} and implies that if the image is conjugate into $\mathcal{N}$, it is D_4. It follows that if $|J_{X_0}(\alpha)| \neq \emptyset$, then either α has type I or $(a,b,c) = (2,4,4)$. Combined with the previous paragraph, this implies that $\Delta(\alpha, \beta) \leq 2$ when α has T-type or O-type or (a,b,c) is either $(2,3,6)$ or $(3,3,3)$.

Suppose then that $(a,b,c) = (2,4,4)$, Proposition 5.3 of [**Bo**] combines with Proposition 5.2(2) to show that

$$\Delta(\alpha, \beta) \leq 1 + \frac{|J_{X_0}(\alpha)|}{s_{X_0}} \leq 1 + 2 = 3$$

Finally suppose that α has I-type. Then $|J_{X_0}(\alpha)| \leq 2$ ([**BZ1**, Lemma 5.3]) and so as Proposition 10.2 shows that

$$\Delta(\alpha, \beta) = 1 + \frac{2|J_{X_0}(\alpha)|}{s_{X_0}}$$

is at most 4, we have $\Delta(\alpha, \beta) \leq 3$.

The remaining case to consider is when $d = 2$ and $n > 2$. In this case we have $s_{X_0} \geq 2$. Further, no representation of $\pi_1(M)$ whose character lies on X_0 has image D_2 so the reader will verify that $2|J_{X_0}(\alpha)| - |\mathcal{N}(\alpha)| \leq 4$ with equality implying that α is of I-type. Proposition 5.2(2) now gives the distance bounds in Proposition 10.7. This completes the proof. $\diamondsuit$

CHAPTER 11

Algebraic and embedded n-gons in X^ϵ

We prove several results which will be used in the remainder of the paper. First we fix some notation.

The twice punctured essential torus F separates M into two components X^ϵ, $\epsilon \in \{+,-\}$. Let b_1, b_2 denote the components of ∂F. These two circles cut ∂M into two annuli B^ϵ, $\epsilon \in \{+,-\}$, such that $\partial X^\epsilon = F \cup B^\epsilon$. Let $\widehat{b}_1$ and $\widehat{b}_2$ be two disjoint meridian disks of the filling solid torus V_β bounded by b_1 and b_2. These two disks cut V_β into two 3-balls H^ϵ, $\epsilon \in \{+,-\}$, such that $\partial H^\epsilon = B^\epsilon \cup \widehat{b}_1 \cup \widehat{b}_2$. Recall that $\widehat{F} = F \cup \widehat{b}_1 \cup \widehat{b}_2$ is an incompressible torus in $M(\beta)$ and let $\widehat{X}^\epsilon = X^\epsilon \cup H^\epsilon$. Here H^ϵ can be considered as an 2-handle attached to X^ϵ along B^ϵ.

Fix a sign ϵ. An *n-gon* D in X^ϵ is a singular disk D with $\partial D \subseteq \partial X^\epsilon = F \cup B^\epsilon$ such that $\partial D \cap B^\epsilon$ is a set of n embedded essential arcs in B^ϵ, called the *corners* of D, and $\partial D \cap F$ is a set of n singular paths, called the *edges* of D. As we go around ∂D in some direction we get a cyclic sequence of $Y^{\pm 1}$ where Y, Y^{-1} indicate that ∂D is running across B^ϵ from b_1 to b_2 or from b_2 to b_1, respectively. In this way D determines an unreduced cyclic word $W = W(Y^{\pm 1})$, well-defined up to inversion. We say that D is an *algebraic m-gon* if the absolute value of the exponent sum of W is m. We say that D is *essential* if the map $(D, \partial D) \to (X^\epsilon, \partial X^\epsilon)$ is essential as a map of pairs. For instance, an algebraic m-gon is essential if $m > 0$.

We use the term *monogon, bigon* or *trigon* for n-gon when $n = 1, 2$ or 3 respectively. We call an n-gon an *algebraic monogon, algebraic bigon* or *algebraic trigon* if it is an algebraic m-gon for $m = 1, 2$ or 3.

LEMMA 11.1. *Suppose that D is an embedded n-gon in X^ϵ for some $n \leq 3$.*

(1) *D is not a monogon.*
(2) *If D is an algebraic n-gon, then $t_1^\epsilon = 0$ and $\widehat{X}^\epsilon$ is Seifert fibred over D^2 with two cone points, one of order n.*
(3) *If $n = 3$ and D is an algebraic monogon, then there is an incompressible separating annulus $(A_\epsilon, \partial A_\epsilon) \subset (X^\epsilon, F)$ which is the frontier of a trefoil knot exterior $Q \subset \widehat{X}^\epsilon$ such that*
   (a) *$Q \cap \widehat{F}$ is an $\widehat{F}$-essential annulus whose slope on ∂Q is the meridional slope of Q;*
   (b) *if k_ϵ is a core arc of the handle H^ϵ, then $(k_\epsilon, \partial k_\epsilon)$ can be isotoped in $(Q, \partial Q)$ to lie as a core arc of an essential vertical annulus of Q.*

PROOF. If D is a monogon, its union with H^ϵ has a 3-ball regular neighbourhood across which we could isotope the essential torus $\widehat{F}$ into M, which is impossible. Thus (1) holds, so that $n = 2$ or 3.

Suppose that the embedded n-gon D is also an algebraic n-gon. Let $e_1, \ldots, e_n \subset F$ be the edges of D. If $\hat{b}_1 \cup \hat{b}_2 \cup e_1 \cup \ldots \cup e_n$ is contained in a 2-disk $D_0 \subset \widehat{F}$, then a regular neighbourhood of $D_0 \cup D \cup H^\epsilon$ in $M(\beta)$ is a once-punctured lens space of order n. But then as $\widehat{F}$ is essential in $M(\beta)$, the latter would be reducible, contrary to Assumption 4.3(6). We claim that $\hat{b}_1 \cup \hat{b}_2 \cup e_1 \cup \ldots \cup e_n$ is contained in an $\widehat{F}$-essential annulus A as depicted in Figure 11.1. This is obvious when $n = 2$. To see it is the case when $n = 3$, we just need to note that each of the edges e_1, e_2, e_3 connects b_1 and b_2 and that if the endpoints of the corners of D at b_1 occur in clockwise (respectively anticlockwise) order, then their other endpoints at b_2 occur in anticlockwise (respectively clockwise) order.

Now define Q to be a regular neighbourhood of $A \cup H^\epsilon \cup D$ in $\widehat{X}^\epsilon$. The reader will verify that $\pi_1(Q) \cong \langle t, x : tx^n \rangle$ where t generates $\pi_1(A)$ and x corresponds to the core of H^ϵ. Thus Q is a solid torus in which A has winding number n. Let A_ϵ be the frontier of Q in X^ϵ. Then A_ϵ is an annulus of winding number n in Q.

Since $\widehat{X}^\epsilon - \text{int}(Q)$ is contained in M and has boundary a torus containing an $\widehat{F}$-essential annulus, it must be also a solid torus. Furthermore A_ϵ has winding number larger than one in $\widehat{X}^\epsilon - \text{int}(Q)$ as otherwise $\widehat{X}^\epsilon$ would be a solid torus and thus $\widehat{F}$ would be compressible. Thus A_ϵ is an essential annulus in X^ϵ. It follows that $\widehat{X}^\epsilon$ is Seifert fibred over D^2 with two cone points, one of order n.

Recall from Chapter 9 that t_1^ϵ is an even positive integer bounded above by 2. If $t_1^\epsilon = 2$, then $\breve{\Phi}_1^\epsilon$ is a collar on ∂F, so there are no essential homotopies in (X^ϵ, F) of large maps to F. Let K_β be the core of the β-filling solid torus in $M(\beta)$ and $k_\epsilon = K_\beta \cap H^\epsilon$. Then k_ϵ is a core arc of H^ϵ and by construction, the pair $(k_\epsilon, \partial k_\epsilon)$ is isotopic in (Q, A) to a transverse arc of A_ϵ. Thus k_ϵ is a transverse arc in an essential annulus A' properly embedded in $\widehat{X}^\epsilon$. Consider a tubular neighbourhood N' of A' containing $H^\epsilon \setminus (\widehat{b}_1 \cup \widehat{b}_2)$ in its interior and set $N_0' = N' \cap M$. Then $N_0' \cap F$ is a disjoint union of two once-punctured tori A_1^0 and A_2^0 and $(N_0', A_1^0 \sqcup A_2^0) \cong (A_1^0 \times I, A_1^0 \times \partial I)$. But then there is an essential homotopy (X^ϵ, F) of the large subsurface A_1^0 of F, a contradiction. Thus $t_1^\epsilon = 0$, which completes the proof of (2).

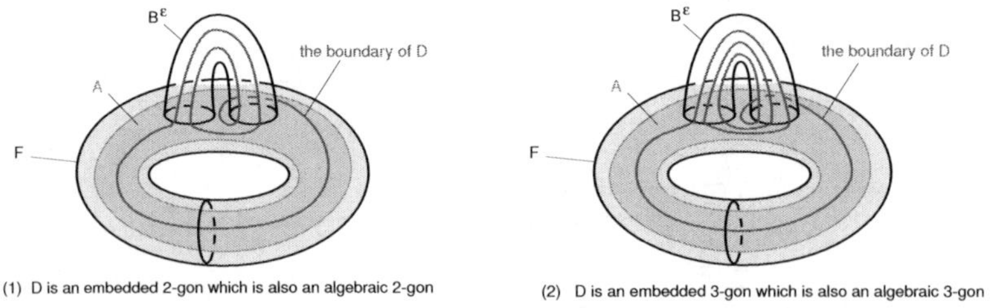

FIGURE 11.1. The boundary of D and the $\widehat{F}$-essential annulus A.

Finally suppose that $n = 3$ and D is an algebraic monogon. Then we may assume that the edges, denoted e_1, e_2, e_3, and the corners, denoted u, v, w, of D are as shown in Figure 11.2(1), where an endpoint of an edge (or a corner) of D is labeled 1 or 2 if it lies in b_1 or b_2. For each of $i = 1, 2$, the 'loop' $\widehat{b}_i \cup e_i$ cannot be contained in a disk in $\widehat{F}$, for otherwise $(D, \partial D) \subset (X^\epsilon, \partial X^\epsilon)$ can be isotoped into an embedded monogon, contradicting part (1) of the lemma. Hence the two loops

$\widehat{b}_1 \cup e_1$ and $\widehat{b}_2 \cup e_2$ are essential and parallel in $\widehat{F}$. So the edges of D are contained in the annulus A shown in Figure 11.2(2).

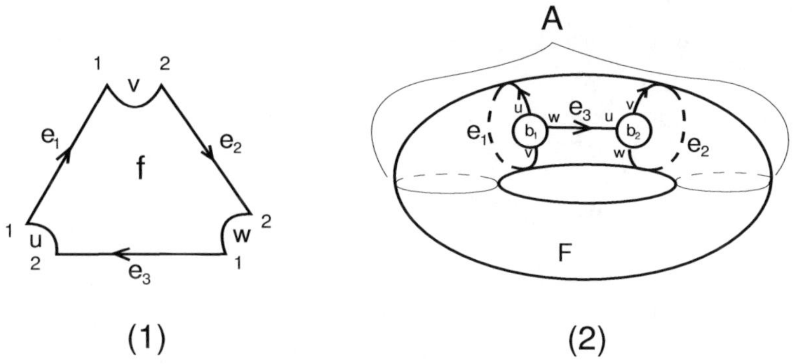

FIGURE 11.2. The case of an embedded 3-gon which is an algebraic monogon.

Now define Q to be a regular neighbourhood of $A \cup H^\epsilon \cup D$ in $\widehat{X}^\epsilon$. Choose a fat base point in $\widehat{F}$ which contains $b_1 \cup b_2 \cup e_3$. Let x be the element of $\pi_1(\widehat{X}^\epsilon)$ represented by an essential arc of the annulus B^ϵ running from b_1 to b_2 and let t be the element represented by the edge e_1 oriented as indicated in Figure 11.2(2). Then Q has the following fundamental group:

$$\pi_1(Q) = \langle t, x : txtx^{-2} \rangle$$

Note that $\pi_1(Q)$ injects into $\pi_1(\widehat{X}^\epsilon)$ since the frontier of Q in $\widehat{X}^\epsilon$ is an incompressible annulus in $(\widehat{X}^\epsilon, \widehat{F})$. Let $y = tx$, then $\pi_1(Q) = \langle x, y : y^2 = x^3 \rangle$ and we see that Q is homeomorphic to the trefoil knot exterior with the loop t as a meridian. In particular the loop t is distance 1 from a regular fibre of Q in ∂Q. A more explicit geometric illustration of the situation is given in part (1) of Figure 11.3, where $A \times I \cup H^\epsilon$ is a genus two handlebody, the brown loop and the green loop bound two disjoint disks which cut the handle body into a 3-ball, the face f is attached to the handlebody along the red loop which intersects the green loop three times with the same sign and intersects the brown curve twice with the same sign.

In Figure 11.3, the blue loop is the boundary of an essential separating disk D_* of the handlebody. The disk D_* separates the handlebody into two solid tori. Let V be the one which contains the green loop. The red loop intersects the blue loop twice (with opposite signs). Let e be the part of the red curve in ∂V and let $E = e \times I$ be a regular neighbourhood of e in $\partial V - \partial D_*$. Observe that $D_* \cup E$ is a vertical essential annulus of Q. Now one can see from Figure 11.3 that the arc $(k_\epsilon, \partial k_\epsilon)$ can be isotoped in $(Q, \partial Q)$ to lie in a position as shown in part (2) of Figure 11.3. This completes the proof. ♢

COROLLARY 11.2. *Suppose that* $t_1^\epsilon = 0$. *If there is a trigon face of* Γ_F *lying in* X^ϵ *which is an algebraic monogon, then* $\widehat{X}^\epsilon$ *is a trefoil exterior.*

PROOF. Suppose that f is a trigon face of Γ_F lying in X^ϵ which is an algebraic monogon. The hypothesis that $t_1^\epsilon = 0$ implies that $\widehat{X}^\epsilon$ admits a Seifert structure with base orbifold a disk with two cone points. We must show that the orders of these cone points are 2 and 3.

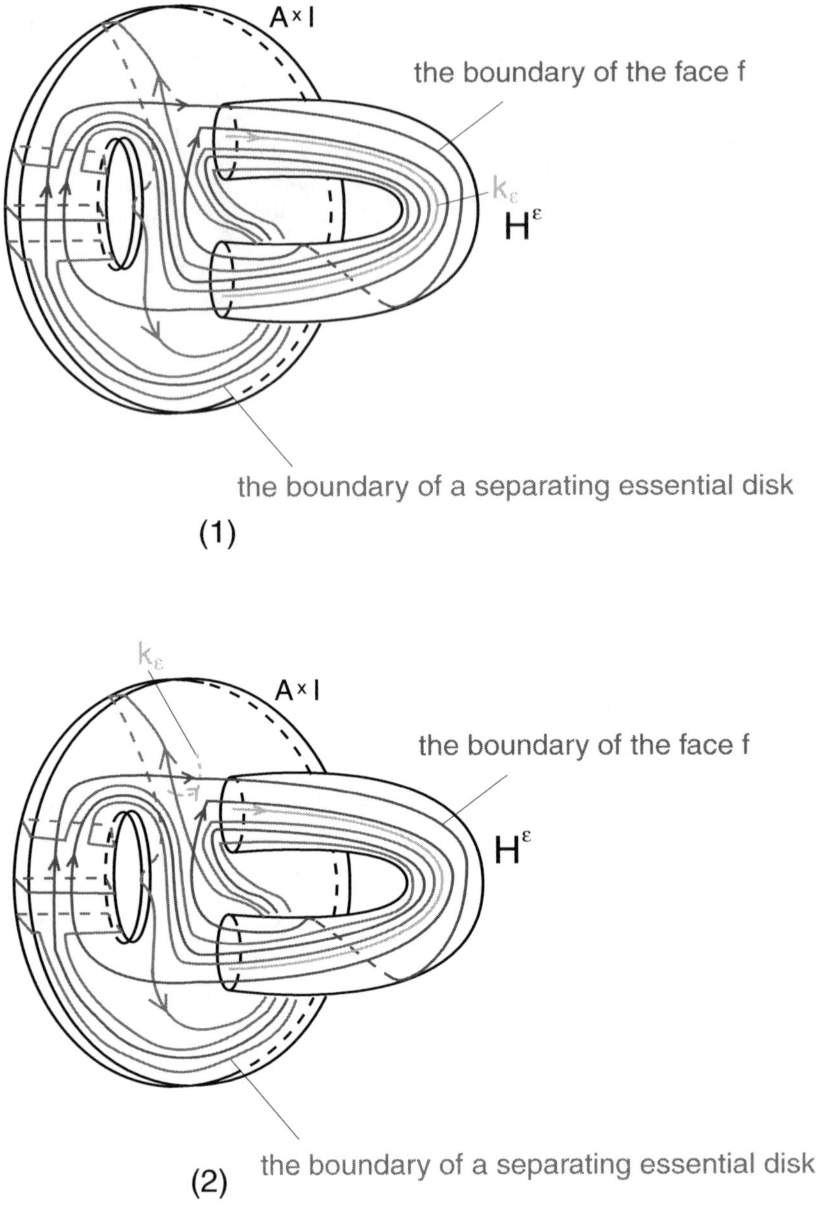

FIGURE 11.3. A geometric illustration of the manifold Q and the arc k_ϵ.

Combining Lemma 11.1(1) and the loop theorem with respect to the (mod 2) intersection with the class of b_1 in $H_1(\partial \widehat{X}^\epsilon)$ ([**He**, Theorem 4.10]), we see that there is an embedded trigon in X^ϵ. If it is an algebraic monogon we are done by Lemma 11.1(3). If it is an algebraic trigon, then the base orbifold of $\widehat{X}^\epsilon$ has a cone point of order 3 by Lemma 11.1(2). Now apply the loop theorem to f with respect to the (mod 3) intersection with the class of b_1. The result is either an embedded trigon which is an algebraic monogon, so we are done, or an embedded bigon which is also an algebraic bigon, in which case the base orbifold of $\widehat{X}^\epsilon$ has a cone point of order 2 (Lemma 11.1(2)), so we are done. $\diamondsuit$

CHAPTER 12

The proof of Theorem 1.2 when F separates but is not a semi-fibre and $t_1^+ + t_1^- > 0$

The goal of this chapter is to prove Theorem 1.2 in the case that F separates, is not a semi-fibre, and $t_1^+ + t_1^- > 0$. More particularly we show:

PROPOSITION 12.1. *Suppose that Assumptions* 4.3 *and* 9.2 *hold, F is separating, and $t_1^+ + t_1^- > 0$. Then $\Delta(\alpha, \beta) \leq 4$.*

We begin with a lemma.

LEMMA 12.2. *Suppose that $t_1^+ + t_1^- > 0$.*

(1) *If $t_1^+ + t_1^- > 2$, then* $\Delta(\alpha, \beta) \leq \begin{cases} 2 & \text{if } M(\alpha) \text{ is very small} \\ 3 & \text{otherwise} \end{cases}$

(2) *Suppose that $t_1^+ + t_1^- = 2$, say $t_1^\epsilon = 2$ and $t_1^{-\epsilon} = 0$.*
   (a) *X^ϵ admits no essential bigons or embedded trigons which are algebraic trigons.*
   (b) $\Delta(\alpha, \beta) \leq \begin{cases} 5 & \text{if } M(\alpha) \text{ is very small} \\ 6 & \text{otherwise} \end{cases}$
   (c) *If* $\Delta(\alpha, \beta) \geq \begin{cases} 4 & \text{if } M(\alpha) \text{ is very small} \\ 5 & \text{otherwise} \end{cases}$, *then $\widehat{X}^\epsilon$ is the union of a trefoil exterior Q and a solid torus U along an annulus $A_\epsilon = \partial Q \cap \partial U$ which is meridional in Q and of winding number 2 or more in U.*

PROOF. Recall from Chapter 9 that t_1^+ and t_1^- are even integers satisfying $0 \leq t_1^+, t_1^- \leq 2$. If $t_1^+ = t_1^- = 2$, then $\breve{\Phi}_1^+ = \breve{\Phi}_1^-$ is a collar on ∂F, so there are no essential homotopies in (M, F) of positive length of large maps to F. Hence no pair of distinct edges in Γ_F are parallel ([**BGZ2**, Proposition 11.3]) and therefore the valency of each vertex of $\overline{\Gamma}_F$ equals that of the vertices of Γ_F, which is $2\Delta(\alpha, \beta)$. If $M(\alpha)$ is very small, $\overline{\Gamma}_F$ is contained in a disk so that there is a vertex of valency at most 5. Hence $\Delta(\alpha, \beta) \leq 2$. Otherwise $\overline{\Gamma}_F$ is contained in a torus and so has a vertex of valency at most 6. Thus $\Delta(\alpha, \beta) \leq 3$, which proves (1).

Assume next that $t_1^+ + t_1^- = 2$. Without loss of generality we can suppose that $t_1^+ = 2$ and $t_1^- = 0$. Since $t_1^+ = 2$, $\breve{\Phi}_1^+$ is a collar on ∂F, so there are no essential homotopies of large maps in (X^+, F). Hence there are no essential bigons in X^+. Lemma 11.1(2) implies that there are no embedded trigons in X^+ which are algebraic trigons, so (2)(a) holds.

Since there are no essential bigons in X^+, all bigon faces of Γ_F lie in X^-. It follows that the weight of each edge of $\overline{\Gamma}_F$ is no more than 2. Since the valency of any vertex of Γ_F is $2\Delta(\alpha, \beta)$ it follows that $\Delta(\alpha, \beta)$ is bounded above by the valency of any vertex v of $\overline{\Gamma}_F$ with equality if and only if the weight of each edge incident to v is 2. When the immersion surface Y containing $\overline{\Gamma}_F$ is a torus (cf.

(9.1)), there is a vertex of $\overline{\Gamma}_F$ of valency 6 or fewer, so $\Delta(\alpha,\beta) \leq 6$. When it is a disk, for instance when $M(\alpha)$ is very small, there is a vertex of $\overline{\Gamma}_F$ of valency 5 or fewer, so $\Delta(\alpha,\beta) \leq 5$. Thus (2)(b) holds.

Now assume that the hypotheses of (2)(c) hold. Then from the previous paragraph we see that for each vertex v of $\overline{\Gamma}_F$,

$$\text{valency}_{\overline{\Gamma}_F}(v) \geq \Delta(\alpha,\beta) \geq \begin{cases} 4 & \text{if } M(\alpha) \text{ is very small} \\ 5 & \text{otherwise} \end{cases} \tag{12.1}$$

Claim 12.3. Γ_F has a trigon face f which lies in X^+.

Proof of Claim 12.3. Let $\varphi_j(v)$ denote the number of corners of j-gons incident to a vertex v of Γ_F and set

$$\mu(v) = \varphi_2(v) + \frac{\varphi_3(v)}{3}$$

If there is a vertex v of Γ_F such that $\mu(v) > 2\Delta(\alpha,\beta) - 4$, then our hypotheses combine with [**BGZ2**, Proposition 12.2 (1)] to show that either v has valency 4 and $\varphi_3(v) \geq 1$ or v has valency 5 and $\varphi_3(v) \geq 4$. In either case, our assumed lower bound on $\Delta(\alpha,\beta)$ implies that $\text{valency}_{\overline{\Gamma}_F}(v) = \Delta(\alpha,\beta)$. But as we remarked above, this implies that each edge of $\overline{\Gamma}_F$ incident to v has weight 2. This implies that each of the faces of $\overline{\Gamma}_F$ incident to v lie in X^+. In particular, the fact that $\varphi_3(v) > 0$ implies that there is a trigon face of Γ_F lying in X^+.

Assume next that $\mu(v) \leq 2\Delta(\alpha,\beta) - 4$ for each v. Then Corollary 12.4 of [**BGZ2**] implies that the immersion surface is a torus and $\mu(v) = 2\Delta(\alpha,\beta) - 4$ for each v. Proposition 12.2 of [**BGZ2**] then implies that each vertex of $\overline{\Gamma}_F$ has valency 5 or 6.

If $\Delta(\alpha,\beta) = 6$ then each vertex has valency 6, each edge of $\overline{\Gamma}_F$ has weight 2, and [**BGZ2**, Proposition 12.2] implies that each face of $\overline{\Gamma}_F$ is a trigon. Thus $\overline{\Gamma}_F$ is a hexagonal triangulation and any of its faces can serve as f.

Finally suppose that $\Delta(\alpha,\beta) = 5$. If there is a vertex v of valency 5 the edges of $\overline{\Gamma}_F$ incident to v have weight 2 and $\varphi_3(v) = 3$ ([**BGZ2**, Proposition 12.2]), so we are done. If all vertices of $\overline{\Gamma}_F$ have valency 6, then all its faces are trigons and each vertex is incident to exactly four edges of weight 2, from which the existence of f follows. This completes the proof of Claim 12.3. ◇

Apply the loop theorem to replace a trigon face f of Γ_F lying in X^+ by an embedded monogon, bigon or trigon $(D, \partial D) \subset (X^+, \partial X^+)$, with each edge of D an essential arc in $(F, \partial F)$. Lemma 11.1(1) shows that D is not a monogon while as we noted above, the condition that $t_1^+ = 2$ implies that it is not a bigon. Thus it is a trigon. Lemma 11.1(2) shows that it is an algebraic monogon.

According to Lemma 11.1(3), there is an incompressible separating annulus $(A_+, \partial A_+) \subset (X^+, F)$ which is the frontier of a trefoil knot exterior $Q \subset \widehat{X}^+$ such that

(a) $Q \cap \widehat{F}$ is an $\widehat{F}$-essential annulus whose slope on ∂Q is the meridional slope of Q;
(b) if k_+ is a core arc of the handle H^+, then $(k_+, \partial k_+)$ can be isotoped in $(Q, \partial Q)$ to lie as
a core arc of an essential vertical annulus of Q.

If A_+ is inessential in (X^+, F), then Q is isotopic to $\widehat{X}^+$ in $M(\beta)$. But then, as in the proof of Lemma 11.1(2), $t_1^+ = 0$, which is a contradiction. Thus A_+ is essential in (X^+, F). On the other hand, $A_+ \cup (F \setminus Q \cap F)$ is a separating torus in $X^+ \subset M$ which contains an $\widehat{F}$-essential annulus. Thus it bounds a solid torus $U \subset X^+$ such that $U \cap Q = A_+$ and A_+ has winding number 2 or more in U. It follows that $\widehat{X}^+$ is the union of a trefoil exterior with a cable space in such a way that the cable slope is identified with the meridional slope of the trefoil. This completes the proof of Lemma 12.2. ◇

LEMMA 12.4. *Suppose that Assumptions 4.3 and 9.2 hold, F separates, $t_1^+ = 2, t_1^- = 0$, and*

$$\Delta(\alpha, \beta) \geq \begin{cases} 4 & \text{if } M(\alpha) \text{ is very small} \\ 5 & \text{otherwise} \end{cases}$$

If f is a trigon face of Γ_F contained in X^+, then there is a homotopy $H : Y \times I \to M(\alpha)$ of h to a new immersion h' satisfying

- *H has support in an arbitrarily small neighbourhood in Y of a compact subset of f disjoint from its corners;*
- *$(h')^{-1}(F) = h^{-1}(F)$;*
- *$h'(f) \subset Q$ where $Q \subset \widehat{X}^+$ is the trefoil exterior described in Lemma 12.2(2).*

PROOF. Lemmas 11.1(3) and 12.2(3) imply that

- $\widehat{X}^+$ is the union of a trefoil exterior Q and a solid torus U along an essential annulus
$$(A_+, \partial A_+) = (\partial Q \cap \partial U, \partial Q \cap \partial U \cap F)$$
which is meridional in ∂Q and of winding number 2 or more in U;
- $Q \cap \widehat{F}$ is an $\widehat{F}$-essential annulus A containing ∂F and $H^+ \subset Q - A_+$;
- $U \cap F$ is an $\widehat{F}$-essential annulus E.

First homotope h, preserving $h^{-1}(F)$, in a small neighbourhood of f in Y disjoint from its corners so that $h|_f$ is transverse to A_+. A loop component of $f \cap h^{-1}(A_+)$ lies in $\text{int}(f) \subset \text{int}(X^+)$ and so as $\pi_2(Q, A_+) = \pi_2(U, A_+) = 0$, any such component could be removed by a homotopy with support in the interior of f. In particular, the homotopy is fixed on $h^{-1}(F)$.

Next suppose there is an arc component of $f \cap h^{-1}(A_+)$ whose end-points lie on an edge of f. Let a be such an arc which is outermost in f and D_0 a disk in f whose boundary consists of a and an arc b in the edge of f containing ∂a. Then $h|_{(D_0, a \cup b)}$ represents an element of $\pi_2(Q, \partial Q)$ or $\pi_2(U, \partial U)$ which is necessarily zero. To see why the latter holds, note that the long exact homotopy sequence of the pair $(Q, \partial Q)$ shows that $\pi_2(Q, \partial Q) = 0$ while that of $(U, \partial U)$ yields a short exact sequence of abelian groups

$$0 \to \pi_2(U, \partial U) \xrightarrow{\partial} \pi_1(\partial U) \to \pi_1(U) \to 0,$$

from which we see that $\pi_2(U, \partial U) \cong \mathbb{Z}$ is generated by a meridional disk of U.

Suppose that $h|_{(D_0, a \cup b)}$ represents a non-zero homotopy class in its group. Then $h(D_0) \subset U$ and the fundamental class of ∂D_0 is sent by h_* to a non-zero multiple of the meridional class in $H_1(\partial U)$. If $h(\partial a)$ is contained in a single component of ∂A_+, then $h|_{D_0}$ is homotopic to a map $(D_0, \partial D_0) \to (U, E)$ and so $h|_{(D_0, a \cup b)}$ lies in the image of $0 = \pi_2(U, E) \to \pi_2(U, \partial U)$, a contradiction. Thus the two points of $\partial a = \partial b$ are sents into distinct components of ∂A_+ by h. It follows that $h|_a$ is homotopic (rel ∂a) to a transverse arc in A_+ and $h|_b$ is homotopic (rel $\partial b = \partial a$) to a transverse arc in the annulus E. But then $h|_{\partial D_0}$ is homotopic to a simple closed curve in ∂U which must be meridional. By construction, this curve has algebraic intersection ± 1 with the core of A_+. On the other hand, the absolute value of this intersection is the winding number of A_+ in U, which is therefore 1, a contradiction.

Suppose next that $h|_{(D_0, a\cup b)}$ represents zero in $\pi_2(W, \partial W)$ where W is Q or U as the case may be. Then we can remove a from $f \cap h^{-1}(A_+)$ by a homotopy preserving $h^{-1}(F)$ with support in an arbitrarily small neighbourhood of D_0 in Y. More precisely, the long exact sequence in homotopy of the pair $(W, \partial W)$ shows that $h|_{\partial D_0}$ is homotopically trivial in ∂W. Hence its algebraic intersection with either component of ∂A_+ is zero. It follows that ∂b is contained in a single component of ∂A_+ and therefore can be homotoped (rel ∂b) into that component. Hence we can homotope h, preserving $h^{-1}(F)$ and with support in an arbitrarily small neighbourhood of b in Y so as to replace a by a closed component of $f \cap h^{-1}(A_+)$ contained in the interior of f. The latter can be removed as in the the first paragraph of this proof.

Finally suppose that there is an arc component of $f \cap h^{-1}(A_+)$ whose end-points lie on distinct edges of f. Choose a cornermost such arc a. Then a cobounds a disk D_0 in f with arcs b, c, d where $h(D_0) \subset Q$, $h(b \cup d) \subset A$, $h(c)$ runs once over the annulus B^+, and the arcs are oriented so that the product $a * b * c * d$ is well-defined and represents a fundamental class of ∂D_0.

Choose the base point for $\pi_1(Q)$ to be $p_0 = h(d \cap a) \in A_+ \cap F \subset A$ and recall the presentation $\pi_1(Q; p_0) = \langle t, x : txtx^{-2} \rangle$ described above. By construction, we can suppose that $t \in \pi_1(A; p_0) \leq \pi_1(Q; p_0)$. On the other hand, x may be assumed to be represented by $h(d)^{-1} * h(c)^{-1} * \gamma$ where γ is a path in A from $h(b \cap c)$ to $h(a \cap d)$. Then the existence of the disk D_0 implies that $h(d)^{-1} * h(c)^{-1} * \gamma \simeq h(a) * h(b) * \gamma$ rel $\{0, 1\}$. But then $x = [h(a) * h(b) * \gamma] \in \pi_1(\partial Q; p_0) \leq \pi_1(Q; p_0)$ which would imply that $\pi_1(\partial Q; p_0) \to \pi_1(Q; p_0)$ is surjective, a contradiction. Thus this case does not arise, which completes the proof of the lemma. ◇

LEMMA 12.5. *Suppose that Assumptions* 4.3 *and* 9.2 *hold,* F *is separating,* $t_1^+ = 2, t_1^- = 0$, *and*

$$\Delta(\alpha, \beta) \geq \begin{cases} 4 & \text{if } M(\alpha) \text{ is very small} \\ 5 & \text{otherwise} \end{cases}$$

Then $\overline{\Gamma}_F$ *contains no pair of trigon faces* f_1, f_2 *which share a positive edge* e *of weight* 2.

PROOF. We use the notation from the previous lemma and note that it implies that we can assume $h(f_1) \cup h(f_2) \subset Q$.

Let e_i be the edge of f_i parallel to e and let R be the bigon of Γ_F cobounded by e_1 and e_2. Since e is a positive edge, R is an algebraic bigon. By construction, $\partial R \subset \partial X^- - E$. Applying the loop theorem with respect to the (mod 3) intersection with b_1 yields an embedded n-gon D with boundary contained in $\partial X^- - E$ where $n \leq 2$ such that D is an algebraic m-gon for some $m \not\equiv 0$ (mod 3). Lemma 11.1(1) shows that $n = 2$ and so $m = 2$ as well.

There is a disk D' properly embedded in H^- which contains $k_- = K_\beta \cap H^-$ as a properly embedded arc and which intersects B^- in the two arcs $D \cap B^-$. Then $D \cup D'$ is a Möbius band properly embedded in $\widehat{X}^-$ which contains k_- as a properly embedded essential arc. It has a solid torus regular neighbourhood, V say, which intersects $\widehat{F}$ in an essential annulus $E' \subset \widehat{F} - E$ whose interior contains ∂F and which has winding number 2 in V. Up to isotopy, we can assume that $\partial E' = \partial A_+$.

The annulus $A_- = \partial V \setminus E'$ is properly embedded and separating in X^- and as $t_1^- = 0$, it cobounds a solid torus $V' \subset X^-$ with E in which A_- has winding number two or more. But then $U \cup V$ would be a Seifert manifold over the 2-disk with two cone points which is contained in the hyperbolic manifold M, which is impossible. Thus $\overline{\Gamma}_F$ contains no pair of trigon faces f_1, f_2 which share a positive edge e. $\diamondsuit$

LEMMA 12.6. *Suppose that Assumptions* 4.3 *and* 9.2 *hold, F is separating, $t_1^+ = 2, t_1^- = 0$, and*

$$\Delta(\alpha, \beta) \geq \begin{cases} 4 & \text{if } M(\alpha) \text{ is very small} \\ 5 & \text{otherwise} \end{cases}$$

Then $\Gamma_{\overline{F}}$ contains no pair of trigon faces f_1, f_2 which share a negative edge e of weight 2.

PROOF. We use the notation of Lemma 12.4 and note that it implies that we can assume $h(f_1) \cup h(f_2) \subset Q$.

Let e_i be the edge of f_i parallel to e and let R be the bigon of Γ_F cobounded by e_1 and e_2. Let σ_i be the essential loop $h|(e_i, \partial e_i) : (e_i, \partial e_i) \to (F, \partial F)$. Without loss of generality we can assume that b_i is the origin of σ_i. We can also assume that the images of σ_1, σ_2 are contained in $\dot{\Phi}_1^-$.

If σ_i is $\widehat{F}$-inessential, then the relation associated to f_i (cf. §15 of [**BGZ2**]) implies that the class of $x \in \pi_1(Q) \leq \pi_1(\widehat{X}^+)$ (cf. the proof of Lemma 11.1(3)) is peripheral (cf. [**BGZ2**, Corollary 15.2]), which is false. Thus σ_1 and σ_2 represent $\widehat{F}$-essential loops and therefore ∂R is an essential loop in $\partial X^- - E$. The loop theorem produces an embedded bigon $(D, \partial D) \subset (X^-, \partial X^-)$. The proof of Lemma 12.5 shows that D is not an algebraic bigon, so intersects F in disjoint loops, one based at b_1 and one at b_2. Since F is incompressible, each of the loops is $\widehat{F}$-essential and thus is isotopic in $\widehat{F}$ into ∂E.

The union of D and an appropriate choice of a product region in B^- between the two arcs of $D \cap B^-$ yields an embedded annulus A^- in X^- whose boundary is contained in A and is isotopic in A to $\partial A = \partial E$. After an isotopy in X^-, we can suppose that $\partial A_- = \partial E$. Since $(k_-, \partial k_-)$ is isotopic in $(\widehat{X}^-, \widehat{F})$ into the chosen product region in B^-, A_- is an essential annulus in X^-.

If A_- is non-separating in $\widehat{X}^-$, then $\widehat{X}^-$ is a twisted I-bundle over the Klein bottle and $A_- \cup A_+$ is an embedded non-separating Klein bottle or torus in the hyperbolic manifold M, which is impossible. But if A_- is separating in $\widehat{X}^-$, it splits the latter into two solid tori in each of which A_- has winding number 2. One of these solid tori, V say, is contained in X^- and has boundary $A_- \cup E$. Then $V \cup U$ is a Seifert manifold over a 2-disk with two cone points contained in M, which is impossible. $\diamondsuit$

PROOF OF PROPOSITION 12.1. Suppose otherwise that $\Delta(\alpha, \beta) \geq 5$ and, without loss of generality, that $t_1^+ = 2, t_1^- = 0$. We saw in the proof of Lemma 12.2(2) that the condition $t_1^+ = 2$ implies that each edge of $\overline{\Gamma}_F$ has weight at most 2 and each vertex has valency at least 5. Further, if v is a vertex of $\overline{\Gamma}_F$, then $\text{valency}_{\overline{\Gamma}_F}(v) \geq \Delta(\alpha, \beta)$ with equality if and only if each edge incident to v has weight 2.

Corollary 12.4 of [**BGZ2**] shows that there is a vertex v of Γ_F for which $\mu(v) \geq 2\Delta(\alpha,\beta) - 4$. Proposition 12.2 of that paper then shows that $5 \leq \text{valency}_{\overline{\Gamma}_F}(v) \leq 6$. Hence if k is the number of weight 2 edges incident to v, then

$$10 \leq 2\Delta(\alpha,\beta) = \text{valency}_{\Gamma_F}(v) = 2k + (\text{valency}_{\overline{\Gamma}_F}(v) - k) \leq 6 + k$$

Hence $k \geq 4$.

If $\mu(v) > 2\Delta(\alpha,\beta) - 4$, Proposition 12.2 of [**BGZ2**] shows that v has valency 5 with at least four trigon faces incident to it, two of which share a weight 2 edge as $k \geq 4$, contrary to Lemmas 12.5 and 12.6.

On the other hand if $\mu(v) = 2\Delta(\alpha,\beta) - 4$, then Proposition 12.2 of [**BGZ2**] shows that v has valency 5 with at least four trigon faces incident to it or 6 with six trigon faces incident to it. In either case the fact that $k \geq 4$ implies that there are a pair of trigons incident to v which share a weight 2 edge, contrary to Lemmas 12.5 and 12.6. This final contradiction shows that $\Delta(\alpha,\beta) \leq 4$. ♢

CHAPTER 13

Background for the proof of Theorem 1.2 when F separates and $t_1^+ = t_1^- = 0$

Throughout the the remainder of the paper we assume that F separates, is not a semi-fibre, and $t_1^+ = t_1^- = 0$.

Recall that when $t_1^+ = t_1^- = 0$, $\widehat{X}^+$ and $\widehat{X}^-$ are Seifert fibred over the disk with two cone points [**BGZ2**, Proposition 7.4]. Further, if $\widehat{X}^\epsilon$ has base orbifold $D^2(p_\epsilon, q_\epsilon)$, $2 \leq p_\epsilon \leq q_\epsilon$, then

- $2 < p_\epsilon \leq q_\epsilon$ if and only if $\dot{\widehat{\Phi}}_1^\epsilon$ is the union of two once-punctured $\widehat{F}$-essential annuli which are vertical in $\widehat{X}^\epsilon$.
- $2 = p_\epsilon < q_\epsilon$ if and only if $\dot{\widehat{\Phi}}_1^\epsilon$ is a twice-punctured $\widehat{F}$-essential annulus which is vertical in $\widehat{X}^\epsilon$.
- $p_\epsilon = q_\epsilon = 2$ if and only if X^ϵ is a twisted I bundle.

Without loss of generality, we assume that $(p_-, q_-) \leq (p_+, q_+)$ lexicographically. That is,

- $p_- \leq p_+$ and if $p_- = p_+$, then $q_- \leq q_+$.

Since F is not a semi-fibre, $(p_+, q_+) \neq (2, 2)$. An analysis of the possible essential tori in $M(\beta)$ then shows that it is not the union of two twisted I-bundles over the Klein bottle (cf. [**BGZ2**, Corollary 7.6]). Further, under the added assumption that X^- is not a twisted I-bundle, $(p_-, q_-) \neq (2, 2)$, if $M(\beta)$ contains a Klein bottle then $M(\beta)$ is Seifert with base orbifold $S^2(p_+, q_+, p_-, q_-)$ where $p_+ = p_- = 2$.

13.1. The case that $M(\beta)$ is Seifert

Let ϕ_ϵ denote the slope on $\widehat{F}$ of a Seifert fibre of $\widehat{X}^\epsilon$ corresponding to the Seifert structure whose base orbifold is $D^2(p_\epsilon, q_\epsilon)$ and set

$$d = \Delta(\phi_+, \phi_-).$$

When $\widehat{X}^-$ is a twisted I-bundle over the Klein bottle we let ϕ'_- denote the slope on $\widehat{F}$ of a Seifert fibre of $\widehat{X}^-$ corresponding to the Seifert structure whose base orbifold is a Möbius band. It is well-known that

$$\Delta(\phi_-, \phi'_-) = 1.$$

Set

$$d' = \Delta(\phi_+, \phi'_-)$$

and orient ϕ_+, ϕ_-, and ϕ'_- so that

$$d = \phi_+ \cdot \phi_- \quad \text{and } d' = \phi_+ \cdot \phi'_-.$$

If $d = 0$ or X^- is a twisted I-bundle and $d' = 0$, then there are Seifert fibre structures on $\widehat{X}^+$ and $\widehat{X}^-$ which piece together to give one on $M(\beta)$.

Conversely, if $M(\beta)$ is a Seifert fibre space, the separating essential torus $\widehat{F}$ cannot be horizontal as this would imply that $M(\beta)$ has non-orientable Euclidean base orbifold $P^2(2,2)$ or the Klein bottle K. In either case, $M(\beta)$ would be a union of two twisted I-bundles over the Klein bottle, contrary to what we noted above. Thus $\widehat{F}$ is vertical and this implies that there are Seifert structures on $\widehat{X}^+$ and $\widehat{X}^-$ which coincide on $\widehat{F}$. It follows that either $\phi_+ = \phi_-$ (i.e. $d = 0$) or $\widehat{X}^-$ is a twisted I-bundle and $\phi_+ = \phi'_-$ (i.e. $d' = 0$).

PROPOSITION 13.1. *Suppose that Assumptions* 4.3 *and* 9.2 *hold, F is separating but not a semi-fibre, and $t_1^+ = t_1^- = 0$. If $d = 0$, then $\Delta(\alpha, \beta) \leq 1$. In particular, Theorem* 1.2 *holds.*

PROOF. When $d = 0$, $M(\beta)$ is Seifert fibred with base orbifold

$$S^2(p_+, q_+, p_-, q_-).$$

Since $(p_+, q_+) \neq (2,2)$, $S^2(p_+, q_+, p_-, q_-)$ is a hyperbolic 2-orbifold and so [**BGZ1**, Theorem 1.7] implies that β is a singular slope of a closed essential surface in M. Consequently, $\Delta(\alpha, \beta) \leq 1$ by Proposition 4.1. ♢

PROPOSITION 13.2. *Suppose that Assumptions* 4.3 *and* 9.2 *hold, F is separating but not a semi-fibre, and $t_1^+ = t_1^- = 0$, $p_- = q_- = 2$ (i.e. X^- is a twisted I-bundle), $d' = 0$, and $\Delta(\alpha, \beta)$ is even. Then $\Delta(\alpha, \beta) \leq 2$. In particular, Theorem* 1.2 *holds.*

PROOF. By assumption, X^- is a twisted I-bundle over a once-punctured Klein bottle and hence there is a 2-fold cover of $\widetilde{M} \to M$ which restricts to the cover $F \times I \to X^-$ on the $-$-side of F and the trivial double cover on the $+$-side of F. Since $m \equiv 2 \pmod 4$ the boundary of $\widetilde{M}$ is connected.

Now β lifts to a slope β' on $\partial \widetilde{M}$ whose associated filling admits a 2-fold cover $\widetilde{M}(\beta') \to M(\beta)$. Since $d' = 0$, $M(\beta)$ admits a Seifert fibre structure with base orbifold $P^2(p_+, q_+)$ and the reader will verify that $\widetilde{M}(\beta')$ admits a Seifert fibre structure with base orbifold $S^2(p_+, q_+, p_+, q_+) \neq S^2(2,2,2,2)$. Hence β' is a singular slope of some closed essential surface $S \subseteq \widetilde{M}$ ([**BGZ1**, Theorem 1.7]).

Since the distance of α to β is even, α also lifts to a slope α' on $\partial \widetilde{M}$ with the associated filling a 2-fold cover of $M(\alpha)$. Hence $\widetilde{M}(\alpha')$ admits a Seifert fibre structure with base orbifold a 2-sphere with three or four cone points.

Suppose that $\Delta(\alpha, \beta) \geq 4$. It's easy to see that the distance between α' and β' is $\Delta(\alpha, \beta)/2 \geq 2$. Hence as β' is a singular slope for S, S is incompressible in $\widetilde{M}(\alpha')$. As $\widetilde{M}$ is hyperbolic, S cannot be a torus and therefore must be horizontal in $\widetilde{M}(\alpha')$. It cannot be separating as the base orbifold of $\widetilde{M}(\alpha')$ is orientable. Thus it is non-separating. But then Proposition 4.1 implies the distance between α' and β' is at most 1, and therefore $\Delta(\alpha, \beta) = 2\Delta(\alpha', \beta') \leq 2$, a contradiction. Thus $\Delta(\alpha, \beta) \leq 3$, and as this distance is even, $\Delta(\alpha, \beta) \leq 2$. ♢

13.2. Edge weights

LEMMA 13.3. *Suppose that $t_1^+ = t_1^- = 0$ and $\Delta(\alpha, \beta) > 3$.*

(1) *If X^- is not a twisted I-bundle, then,*
   (a) *the edges of $\overline{\Gamma}_F$ have weight at most* 3 *and a negative edge of weight* 3 *determines an $\widehat{F}$-inessential loop;*
   (b) *if there is an edge of weight* 3 *then $p_+ = p_- = 2$;*

(c) *a negative edge of* $\overline{\Gamma}_F$ *incident to a trigon has weight at most* 2, *and if* 2, *then* $d = 1$.

(2) *If* X^- *is a twisted* I*-bundle, then the faces of* $\overline{\Gamma}_F$ *which lie in* X^- *can be assumed to be bigons and its edges can be assumed to have even weight. Further,*

(a) *the edges of* $\overline{\Gamma}_F$ *have weight at most* 4;

(b) *a negative edge of* $\overline{\Gamma}_F$ *incident to a trigon has weight at most* 2.

PROOF OF LEMMA 13.3.

(1)(a) Proposition 13.1 shows that $d \geq 1$, so when X^- is not a twisted I-bundle, $\dot{\Phi}_2^+ \cong \dot{\Phi}_2^- \cong \dot{\Phi}_1^+ \wedge \dot{\Phi}_1^-$ consists of tight components ([**BGZ2**, Lemma 9.2]). Hence $\dot{\Phi}_3^+ = \dot{\Phi}_3^- = \emptyset$ ([**BGZ2**, Proposition 9.3]), so the weight of each edge is at most 3. Further, any negative edge of weight 3 represents loop contained in a tight component of $\dot{\Phi}_2^\epsilon$ for some ϵ, so is $\widehat{F}$-inessential. This proves (1)(a).

(1)(b) Suppose that $p_\epsilon > 2$. Then $\dot{\Phi}_1^\epsilon$ consists of two once-punctured annuli, so each edge of Γ_F which lies in $\dot{\Phi}_1^\epsilon$ is negative and $\widehat{F}$-essential. Thus if an edge $\bar{e}$ of $\overline{\Gamma}_F$ has weight 3, it is negative and its lead edge is $\widehat{F}$-essential, contrary to (1)(a). Thus $p_+ = p_- = 2$.

(1)(c) Let $\bar{e}$ be a negative edge of $\overline{\Gamma}_F$ incident to a trigon f lying in X^ϵ and denote by e_1 the edge of f parallel to $\bar{e}$. Then [**BGZ2**, Corollary 15.2] shows that the loop in $\widehat{F}$ corresponding to e_1 is $\widehat{F}$-essential. If the weight of $\bar{e}$ is 3, $\tau_{-\epsilon}(h(e_1))$ is an $\widehat{F}$-essential loop contained in $\dot{\Phi}_1^+$ and in $\dot{\Phi}_1^-$. But then $\phi_+ = \phi_-$, contrary to the assumption that $d \neq 0$. Hence the weight of $\bar{e}$ can be at most 2, which proves the first part of (1)(c).

(1)(c) Suppose for the second part that $\bar{e}$ has weight 2 and recall from the proof of Lemma 11.1(3) that $\pi_1(\widehat{X}^\epsilon)$ is generated by $\pi_1(\widehat{F})$ and a class x represented by a loop which is the concatenation of an essential arc in B^ϵ and a path in F. Let γ denote the class of the loop $h(e_1)$ in $\pi_1(\widehat{X}^\epsilon)$. Since $X^{-\epsilon}$ is not a twisted I-bundle and the weight of $\bar{e}$ is 2, $\gamma = \phi_{-\epsilon}^k$ for some integer k. The relation associated to f (cf. §15 of [**BGZ2**]) shows that the image x in $\pi_1(\widehat{X}^\epsilon)/\langle\langle\gamma\rangle\rangle = \pi_1(\widehat{X}^\epsilon)/\langle\langle\phi_{-\epsilon}^k\rangle\rangle$ is contained in the image of $\pi_1(\widehat{F})$, so $\pi_1(\widehat{X}^\epsilon)/\langle\langle\phi_{-\epsilon}^k\rangle\rangle$ is abelian. On the other hand, $\pi_1(\widehat{X}^\epsilon)/\langle\langle\phi_{-\epsilon}^k\rangle\rangle$ admits the presentation $\langle u, v : u^{p_\epsilon}, v^{q_\epsilon}, (uv)^{kd}\rangle$ which is abelian only if $kd = \pm 1$. Thus $k = \pm 1$ and $d = 1$.

(2) If X^- is a twisted I-bundle, its base is a once-punctured Klein bottle K_0. We use K_0 to construct an intersection graph Γ_{K_0} as in §2 of [**BCSZ1**]. Though surfaces were assumed orientable in [**BCSZ1**], the construction goes through in our context to yield a graph Γ_{K_0} such that the image of each of its edges in K_0 is an essential path in $(K_0, \partial K_0)$. There is an associated intersection graph Γ_F obtained by doubling each edge of Γ_{K_0}. In this case, each face of Γ_F contained in X^- is a bigon and each edge of $\overline{\Gamma}_F$ has even weight.

(2)(a) If the weight of an edge $\bar{e}$ is larger than 4, then [**BGZ2**, Corollary 11.4 and Proposition 9.4(3)] show that the weight of $\bar{e}$ is 6, $\dot{\Phi}_1^+$ is an $\widehat{F}$-essential twice-punctured annulus, $\dot{\Phi}_3^+$ is the union of two $\widehat{F}$-essential once-punctured annuli, and $\phi_+ = \phi'_-$ so that $M(\beta)$ is Seifert fibred with base orbifold $P^2(2, n)$ for some $n \geq 3$. Let $e_1, \ldots, e_6$ be the parallel successive edges represented by $\bar{e}$. Since e_2 lies in $\dot{\Phi}_4^+ = \dot{\Phi}_3^+$, $\bar{e}$ is a negative edge representing an $\widehat{F}$-essential loop.

Let R be the bigon face between e_3 and e_4, which lie in $\dot{\Phi}_1^+$, and apply the loop theorem to the singular disk $h|_R : (R, \partial R) \to (X^-, \partial X^-)$ to obtain an embedded disk $(D, \partial D) \subset (X^-, \partial X^-)$, which must be a bigon whose two edges d_1 and d_2 are contained in $\dot{\Phi}_1^+$ and two corners of D are essential arcs in the annulus B^-.

If D is an algebraic bigon, then the union of D and a product region in B^- between the two corners of D yields an embedded Möbius band in X^- whose boundary has slope ϕ_- (since the boundary of any embedded Möbius band in $\widehat{X}^-$ has slope ϕ_-). But the boundary of this Möbius band is contained in $\dot{\Phi}_1^+$ and thus has slope $\phi_+ = \phi'_-$. But then $\phi_- = \phi'_-$, which is false.

So D is an algebraic 0-gon. In this case, the union of D and an appropriate choice of a product region in B^- between the two corners of D yields an embedded essential annulus A^- in X^- whose boundary is contained in $\dot{\Phi}_1^+$ and is isotopic to the boundary of the annulus $\widehat{\dot{\Phi}}_1^+$. Since the boundary slope of A^- in $\widehat{F}$ is $\phi_+ = \phi'_-$, A^- is a non-separating annulus in $\widehat{X}^-$. On the other hand we know that the boundary of $\widehat{\dot{\Phi}}_1^+$ bounds a separating essential annulus A^+ in X^+. So up to isotopy the two annuli A^- and A^+ can be pieced together to form a Klein bottle in M, which contradicts the fact that M is hyperbolic.

(2)(b) It follows as in the proof of [**BGZ2**, Lemma 19.6]. ◇

DEFINITION 13.4. We say that an edge of $\overline{\Gamma}_F$ has *maximal weight* if its weight is 3 when X^- is not a twisted I-bundle and 4 when X^- is a twisted I-bundle.

13.3. The trace of an essential annulus in a face of Γ_F

In this section we assume that X^ϵ is not a twisted I-bundle. In particular,

$$q_\epsilon \geq 3.$$

Up to isotopy, there are two choices for an essential annulus A properly embedded in (X^ϵ, F) which cobounds a solid torus in X^ϵ with an $\widehat{F}$-essential annulus in F, one for which A has winding number $q_\epsilon \geq 3$ in the solid torus and one for which it has winding number $p_\epsilon \geq 2$. Throughout the rest of the paper we use

- A_ϵ to denote the essential annulus of winding number q_ϵ, U to denote the associated solid torus, $E = U \cap F$ to denote the associated $\widehat{F}$-essential annulus, and c_1, c_2 to denote the boundary components of A_ϵ;
- A'_ϵ to denote the essential annulus of winding number p_ϵ, V to denote the associated solid torus, $G = V \cap F$ to denote the associated $\widehat{F}$-essential annulus, and c'_1, c'_2 to denote the boundary components of A'_ϵ.

We can write $X^\epsilon = U \cup P \cup V$ where P is the product of a once-punctured annulus with an interval and P has frontier $A_\epsilon \cup A'_\epsilon$ in X^ϵ. Up to isotopy we have

$$(\dot{\Sigma}_1^\epsilon, \dot{\Phi}_1^\epsilon) = \begin{cases} (P, P \cap F) & \text{if } p_\epsilon > 2 \\ (P \cup V, (P \cup V) \cap F) & \text{if } p_\epsilon = 2. \end{cases}$$

Note that

$$(\widehat{P}, A_\epsilon, A'_\epsilon) \cong (A_\epsilon \times I, A_\epsilon \times \{0\}, A_\epsilon \times \{1\})$$

in such a way that P is the exterior of $a_0 \times \{\frac{1}{2}\}$ where a_0 is a transverse arc of A_ϵ.

Index the boundary components of A_ϵ and A'_ϵ so that b_i, c_i, and c'_i lie in the same component of $P \cap F$.

It is clear that $W = V \cup P = X^\epsilon \setminus U$ is a genus 2 handlebody and $\widehat{W} = W \cup H^\epsilon$ is a solid torus in which $A_\epsilon \subset \partial \widehat{W}$ has winding number p_ϵ. Then

$$(\widehat{\dot{\Sigma}_1^\epsilon}, \widehat{\dot{\Phi}_1^\epsilon}) = \begin{cases} (\widehat{P}, \widehat{P} \cap \widehat{F}) & \text{if } p_\epsilon > 2 \\ \\ (\widehat{W}, \widehat{W} \cap \widehat{F}) & \text{if } p_\epsilon = 2. \end{cases}$$

In either case, $(\widehat{P}, \widehat{P} \cap \widehat{F})$ is a regular neighbourhood of A_ϵ in $(\widehat{W}, \widehat{W} \cap \widehat{F})$.

Recall the immersion h of a disk or torus Y into $M(\alpha)$ which yields the graph Γ_F. Fix a face f of Γ_F lying in X^ϵ. We can homotope h to a new immersion such that

- the homotopy has support in an arbitrarily small neighbourhood in Y of a compact subset of f disjoint from its corners;
- the homotopy preserves $h^{-1}(F), h^{-1}(X^+), h^{-1}(X^-)$ and therefore Γ_F;
- the restriction of the new immersion to f is transverse to A_ϵ and further, $f \cap h^{-1}(A_\epsilon)$ has the minimal number of components among all such immersions.

We say that f is in *minimal position* if $h|_f$ satisfies the last of these conditions and assume that this is the case for the remainder of this section.

Label an endpoint of an arc component of $h^{-1}(A_\epsilon) \cap f$ by i if its image under h lies in c_i.

LEMMA 13.5. *Suppose that X^ϵ is not a twisted I-bundle and that f is a minimally positioned face of Γ_F lying in X^ϵ.*

(1) *There are no closed components of $f \cap h^{-1}(A_\epsilon)$.*
(2) *No arc component of $f \cap h^{-1}(A_\epsilon)$ has both of its endpoints lying on the same edge of f.*
(3) *If e is an edge of f and there are endpoints of arcs of $f \cap h^{-1}(A_\epsilon)$ which are successive on $e \cap h^{-1}(A_\epsilon)$ and have the same label, then they cobound a subarc e_0 of e which is mapped by h into $F - E \subset W$.*

PROOF OF LEMMA 13.5.

(1f) Fix an innermost closed component c of $f \cap h^{-1}(A_\epsilon)$ and let $f_0 \subset f$ be the disk which it bounds. Then $h|f_0$ represents an element of $\pi_2(U, A_\epsilon)$ or of $\pi_2(W, A_\epsilon)$ depending on which side of A_ϵ $h(f_0)$ lies on. Since the homomorphisms $\pi_1(A_\epsilon) \to \pi_1(U), \pi_1(A_\epsilon) \to \pi_1(W)$ are injective and $\pi_2(U) \cong \pi_2(W) \cong \{0\}$, the groups $\pi_2(U, A_\epsilon)$ and $\pi_2(W, A_\epsilon)$ are both $\{0\}$. Hence we can homotope h with support in an arbitrarily small neighbourhood of f_0 to eliminate c. The minimality condition on $h|f$ implies that there can be no such component c of $f \cap h^{-1}(A_\epsilon)$. Thus $f \cap h^{-1}(A_\epsilon)$ contains no loop components.

(2f) Since A_ϵ has winding number 2 or more in U and in $\widehat{W}$, the result follows as in the proof of a similar situation described in the proof of Lemma 12.4.

(3f) Suppose that e is an edge of f and there are two arc components of $f \cap h^{-1}(A_\epsilon)$ whose endpoints are successive on e and have the same label. Let e_0 be the subarc of e bounded by these endpoints. Then $h(e_0)$ is a path contained in the annulus E or $F \setminus E$. If $h(e_0)$ is contained in E, it is disjoint from one of the components of $c_1 \cup c_2 = \partial E$. Using the fact that E strong deformation retracts onto either of its boundary components, we could homotope $h|_f$ with support in an arbitrarily small neighbourhood of e_0 to eliminate the two endpoints from $e \cap h^{-1}(A_\epsilon)$. Minimality shows that such a situation does not arise. $\diamondsuit$

We know from Lemma 13.5 that the components of $f \cap h^{-1}(A_\epsilon)$ are arcs connecting different edges of f. Call an arc component of $f \cap h^{-1}(A_\epsilon)$ a *corner-arc* if it is parallel to a corner of f, i.e. the arc joins adjacent edges of f. We call an arc component of $f \cap h^{-1}(A_\epsilon)$ a *cross-arc* if it joins edges of f which are not adjacent.

Since A_ϵ separates X^ϵ into two components, the arcs of $f \cap h^{-1}(A_\epsilon)$ decompose f into subsurfaces, called *tiles*, each of which is mapped by h into U or W. Adjacent tiles lie on different sides of A_ϵ. Colour a tile red if it is mapped into U by h and call it an n-gon if its boundary contains n arcs from $f \cap h^{-1}(A_\epsilon)$.

COROLLARY 13.6. *Suppose that X^ϵ is not a twisted I-bundle and that f is a minimally positioned face of Γ_F lying in X^ϵ. If the intersection of a red tile with an edge e of f is non-empty, the intersection is a closed arc e_0 contained in the interior of e. Further, the labels at the endpoints of e_0 are different.* ♢

LEMMA 13.7. *Suppose that X^ϵ is not a twisted I-bundle and that f is a minimally positioned face of Γ_F lying in X^ϵ.*

(1) *If a is a corner-arc of $f \cap h^{-1}(A_\epsilon)$, then its endpoints have different labels.*
(2) *No two corner-arcs in $f \cap h^{-1}(A_\epsilon)$ are parallel in f. Equivalently, there are no red tile bigons which intersect adjacent edges of ∂f. In particular, a trigon face of Γ_F lying in X^ϵ contains no red tile bigons.*

PROOF.

(1) If the endpoints of a have the same label, say 1, then a can be homotoped, with its endpoints fixed, into $c_1 \subset F$. So in turn, the corner of f parallel to a can be homotoped, with its endpoints fixed, into F, which is impossible.

(2) Suppose otherwise and choose a_1, a_2 from among the arcs of $f \cap h^{-1}(A_\epsilon)$ parallel to a given corner such that $a_1 \cup a_2$ lie in the boundary of a red tile D. By part (1), the endpoints of each a_i have different labels and by Lemma 13.5(3) the labels of endpoints of a_1 and a_2 alternate $1, 2, 1, 2$ around ∂D. This implies that the algebraic intersection in ∂U between the loop $h|_{\partial D}$ and the circle c_1 is 2. Now applying the loop theorem to the singular disk $h|_D$ in U, we get an emdedded disk D_* in U such that the geometric intersection number in ∂U between ∂D_* and c_1 is either 1 or 2, which in turn implies that A_ϵ has winding number 1 or $2 < q_\epsilon$ in U, a contradiction. This completes the proof. ♢

COROLLARY 13.8. *Suppose that X^ϵ is not a twisted I-bundle and that f is a minimally positioned trigon face of Γ_F lying in X^ϵ. Then f contains at most one red tile and if one, it is a trigon. If there is one, then $q_\epsilon = 3$.*

PROOF. Suppose that f contains a red tile. Lemma 13.7(2) shows that it cannot be a bigon, so it is a trigon. Then ∂D consists of three corner-arcs a_1, a_2, a_3, one for each corner of f, and three subarcs e_1^0, e_2^0, e_3^0 of ∂f, one for each boundary edge of f. The labels of endpoints of a_1, a_2, a_3 alternate $1, 2, 1, 2, 1, 2$ around ∂D and so as $h|_D : (D, \partial D) \to (U, \partial U)$, the algebraic intersection in ∂U between the loop $h|_{\partial D}$ and c_1 is 3, which in turn by the loop theorem (as in the proof of Lemma 13.7(2)) implies that q_ϵ, the winding number of A_ϵ in U, is 3. ♢

COROLLARY 13.9. *Suppose that X^ϵ is not a twisted I-bundle and that f is a minimally positioned quad face of Γ_F lying in X^ϵ. If e_1, e_2 are two adjacent edges of f and two or more red tiles are incident to e_1, then there is at most one red tile incident to e_2.*

PROOF. No red tile bigon connects adjacent edges by Lemma 13.7(2) and so as there are at least two red tiles incident to e_1, each of their boundaries contains a cross-arc. It follows that if there is a red tile incident to e_2, it must be a trigon with boundary arcs connecting e_1, e_2 and the other edge e_3 of f adjacent to e_2. But then any other red tile incident to e_2 would have to be a bigon, which we have already ruled out. $\diamondsuit$

LEMMA 13.10. *Suppose that X^ϵ is not a twisted I-bundle and that f is a minimally positioned face of Γ_F lying in X^ϵ. Let $\{a_1, \dots, a_n\}$ be any set of parallel adjacent cross-arcs of $f \cap h^{-1}(A_\epsilon)$.*

(1) *Every a_i has the same label at its endpoints or every a_i has different labels at its endpoints.*
(2) *If the a_i have different labels at their endpoints, then $n \leq 2$ and if 2, the two arcs are contained in different red tiles.*
(3) *If e is an edge of Γ_F incident to f which has maximal weight in $\overline{\Gamma}_F$, the a_i connect the two edges of f adjacent to e, and the labels at the endpoints of a_i are the same, then there is a red tile incident to e.*

PROOF OF LEMMA 13.10.

(1) Since A_ϵ is an essential annulus in X^ϵ and $\pi_2(X^\epsilon, F) = 0$, a properly embedded arc in A_ϵ is essential as a map of pairs to $(A_\epsilon, \partial A_\epsilon)$ if and only if it is essential as a map of pairs to (X^ϵ, F). It follows that for each i, the map $h_i = h| : (a_i, \partial a_i) \to (X^\epsilon, F)$ is essential as a map of pairs if and only if the labels at the endpoints of a_i are distinct. Since any two of the h_i are homotopic as maps to (X^ϵ, F), the first assertion of the lemma follows.

(2) For the second assertion suppose $n \geq 2$ and there is a pair of adjacent arcs a_i and a_{i+1} which lie in the boundary of some red tile D, necessarily a bigon. It follows from Lemma 13.5(3) that the labels of the endpoints of a_i and a_{i+1} alternate around ∂D as $1, 2, 1, 2$, which implies that D is an essential disk in the solid torus U whose algebraic intersection number with c_1 is 2. But then c_1 has winding number 2 in U, a contradiction.

(3) Suppose that the a_i closest to e is a_1 and that there are no red tiles incident to e. Then a_1 cobounds a disk D in f with e, the two corners incident to e, and two subarcs e_0, e_0' of the edges of Γ_F incident to e. Since the labels at the endpoints of a_1 are the same, we can homotope $h|_{e_0 * a_1 * e_0'}$ (rel ∂) to have image in F. But then the disk D provides an essential homotopy of e in X^ϵ contrary to the fact that it has maximal weight. Thus there must be a red tile incident to e. $\diamondsuit$

CHAPTER 14

Recognizing the figure eight knot exterior

In this chapter we describe how work of Martelli, Petronio and Roukema can be used to recognize when M is the figure eight knot exterior. This will be used below to handle the proof of Theorem 1.2 in the case that $d = 1$.

14.1. Exceptional Dehn fillings of the minimally twisted chain link

In [**MP**], Martelli and Petronio classified the non-hyperbolic Dehn fillings of the "magic manifold" M_3, which is the exterior of the hyperbolic chain link of three components in S^3. (This link appears as 6_1^3 in Appendix C of Rolfsen's book [**Rlf**].) We remark that M_3 is the manifold N of the first two chapters of the paper. We've changed notation for convenience.

The following result is contained in [**MP**, Corollary A.6].

PROPOSITION 14.1 (Martelli, Petronio). *Let M be a hyperbolic knot manifold obtained by Dehn filling M_3 along two of its boundary components. If $M(\beta)$ is a toroidal manifold and $M(\alpha)$ is a Seifert fibred manifold such that $\Delta(\alpha, \beta) > 5$, then M is the figure eight knot exterior.* ◇

In [**Rou**], Roukema classified all non-hyperbolic Dehn fillings on the link exterior M_5 of the minimally twisted chain link of five components in S^3 (shown in Figure 14.1).

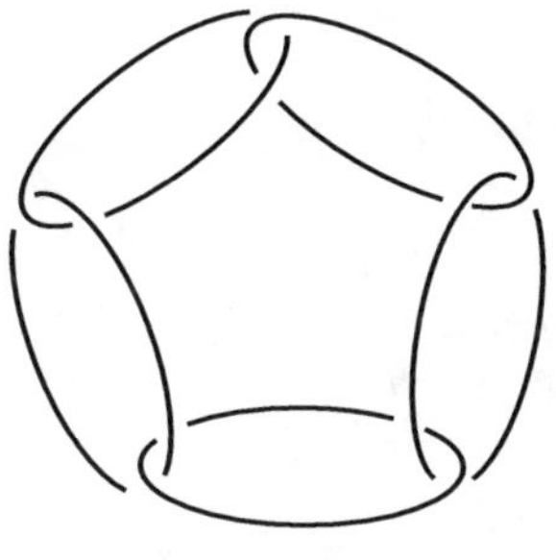

FIGURE 14.1. The minimally twisted 5-chain link

In particular, the following result is contained in [**Rou**, Theorems 3 and 4].

PROPOSITION 14.2 (Roukema). *Suppose that M is a hyperbolic knot manifold obtained by Dehn filling M_5 along four of its boundary components but M cannot be obtained by Dehn filling M_3 along two of its boundary components. Then $\Delta(\alpha, \beta) \leq 4$ for any two non-hyperbolic filling slopes α, β of M.* ◇

Combining the latter two propositions, one gets Corollary 14.3

COROLLARY 14.3. *Suppose that M is a hyperbolic knot manifold obtained by Dehn filling M_5 along four of its boundary components. If $M(\beta)$ is a toroidal manifold and $M(\alpha)$ is a Seifert fibred manifold such that $\Delta(\alpha, \beta) > 5$, then M is the figure eight knot exterior.* ♢

14.2. Essential annuli in $\widehat{X}^\epsilon$ and the figure eight knot exterior

We prove two lemmas which provide sufficient topological conditions for M to be the figure eight knot exterior.

Let K_β be the core of the filling solid torus in forming $M(\beta)$. We noted above that for each ϵ, $\widehat{X}^\epsilon$ admits a Seifert structure over $D(p_\epsilon, q_\epsilon)$. The (I, S^0)-bundle pair $(\Sigma_1^\epsilon, \Phi_1^\epsilon)$ extends to an (I, S^0)-bundle pair $(\widehat{\Sigma_1^\epsilon}, \widehat{\Phi_1^\epsilon})$ in which $k_\epsilon = K_\beta \cap \dot{\widehat{\Sigma_1^\epsilon}}$ is an I-fibre.

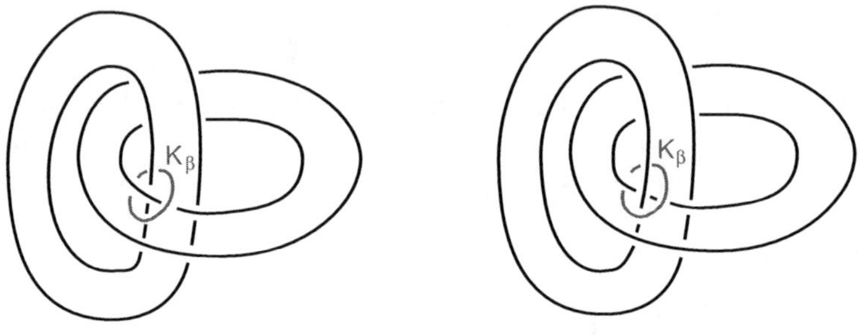

(1) (2)

FIGURE 14.2. Part (1) shows the 5-component link in S^3 resulting from the argument of Lemma 14.4 whose exterior is homeomorphic to that of the link in S^3 shown in Part (2).

LEMMA 14.4. *Suppose that $\Delta(\alpha, \beta) > 5$ and that for each ϵ there is an embedded essential annulus A^ϵ in $\widehat{X}^\epsilon$ such that*

- *∂A^ϵ has slope ϕ_ϵ in $\widehat{F}$;*
- *k_ϵ is an essential arc in A^ϵ;*
- *each boundary component of A^+ intersects each boundary component of A^- transversely and exactly once.*

Then M can be obtained by Dehn filling four boundary components of the minimally twisted 5-chain link exterior and thus it is the figure eight knot exterior.

PROOF. Note that A^ϵ separates $\widehat{X}^\epsilon$ into two solid tori in which A^ϵ has winding numbers p_ϵ, q_ϵ respectively and the core circles of the two solid tori are singular fibres of the Seifert fibred space $\widehat{X}^\epsilon$. Let Z^ϵ be the exterior of the singular fibres in $\widehat{X}^\epsilon$, and let Z be the connected submanifold of $M(\beta)$ which is the union of Z^+ and Z^- in $M(\beta)$ meeting along $\widehat{F}$. Then Z^ϵ is homeomorphic to $Q^\epsilon \times S^1$ where Q^ϵ is a twice punctured disk. Call $\widehat{F}$ the outer boundary component of Z^ϵ and called the other two boundary components of Z^ϵ inner components. Correspondingly the component of ∂Q^ϵ contained in $\widehat{F}$ is called the outer component and the other two components of ∂Q^ϵ are call inner components. The choice of Q^ϵ in the product structure of $Z^\epsilon = Q^\epsilon \times S^1$ is not unique. By Dehn twisting along a vertical annulus in Z^ϵ disjoint from A^ϵ which connects $\widehat{F}$ to an inner boundary component of Z^ϵ, we may assume that Q^+ has been chosen so that its outer boundary component has slope ϕ_- in $\widehat{F}$ and similarly we may assume that Q^- has been chosen so that its outer boundary component has slope ϕ_+ in $\widehat{F}$. It follows that Z is homeomorphic to the exterior of "the untwisted double Hopf link" in S^3 such that the torus $\widehat{F}$ becomes a torus in S^3 which bounds a trivial solid torus V^ϵ in S^3 on each side of $\widehat{F}$ and which separates the two pairs of parallel components of the double Hopf link. Furthermore the twice punctured disk Q^ϵ in Z^ϵ caps off in V^ϵ to a standard meridian disk of V^ϵ. The annulus A^ϵ is a vertical essential annulus in Z^ϵ and it separates the two inner components of ∂Z^ϵ. In fact $A^\epsilon = \delta_\epsilon \times S^1$ where δ_ϵ is a proper arc in Q^ϵ which separates the two inner components of ∂Q^ϵ. The arc k_ϵ in A^ϵ is an essential arc. By Dehn twisting along a vertical annulus $\sigma_\epsilon \times S^1$ in Q^ϵ where σ_ϵ is a proper arc in Q^ϵ connecting the two inner boundary components of Q^ϵ and intersecting δ_ϵ exactly once, we may assume that $K_\beta = k_+ \cup k_-$ is as shown in part (1) of Figure 14.2 (cf. the red coloured component). Let Y be the exterior of K_β in Z, i.e. Y is the exterior of the 5-component link shown in Figure 14.2 (1). Considering Dehn surgery on the component K_β (the red coloured component) in S^3, we see that Y is homeomorphic to the exterior of the 5-component link in S^3 shown in part (2) of Figure 14.2. The latter link is the minimally twisted chain link of 5 components. $\diamondsuit$

REMARK 14.5 (This argument was used in [**GL2**]).

For the exterior N of a link L in S^3 with components $K_0, K_1, \cdots, K_{n-1}$, let $T_0, T_1, \cdots, T_{n-1}$ be the corresponding ordered boundary tori of N. A slope on each T_i will be expressed as (p, q) with respect to the standard (meridian, longitude) coordinates, where p, q are relative prime integers and $q \geq 0$, e.g. $(1, 0)$ is the meridional slope, $(0, 1)$ is the canonical longitude. Let $(0, 0)$ denote the empty slope. Dehn filling of N will be denoted by $N[(p_0, q_0), (p_1, q_1), \cdots, (p_{n-1}, q_{n-1})]$, meaning each T_i is assigned the slope (p_i, q_i), possibly empty. For example when $n = 3$, $N[(0, 0), (-2, 1), (2, 3)]$ means leaving T_0 unfilled and filling T_1, T_2 with slopes $(-2, 1), (2, 3)$ respectively. When N is clear, we may simply denote the filling by $[(p_0, q_0), (p_1, q_1), \cdots, (p_{n-1}, q_{n-1})]$. For each k, $1 \leq k \leq n$, a k-filling of N means filling N along k boundary tori with nonempty slopes and leaving the rest of the boundary tori unfilled. For example $N[(0, 0), (-2, 1), (2, 3)]$ is a 2-filling of N.

LEMMA 14.6. *Suppose that $\Delta(\alpha, \beta) > 5$, X^- is a twisted I-bundle and $d = 1$. Suppose that for each ϵ there is an embedded essential annulus A^ϵ in $\widehat{X}^\epsilon$ such that*

- *∂A^+ has slope ϕ_+ in $\widehat{F}$ and ∂A^- has slope ϕ'_- in $\widehat{F}$;*
- *k_ϵ is an essential arc in A^ϵ;*
- *each boundary component of A^+ intersects each boundary component of A^- transversely and exactly once.*

Then M can be obtained by Dehn filling the boundary components $T_1, \cdots T_6$ of the exterior of the 7-component link $\{K_0, K_1, \cdots, K_6\}$ in S^3 shown in Figure 14.4(2) *such that the slopes on T_3 and T_4 are both $(-1, 1)$. Furthermore M is the figure eight knot exterior.*

PROOF. As in the proof of Lemma 14.4, let Z^ϵ be the exterior of the singular fibres in $\widehat{X}^\epsilon$ with respect to the Seifert structure whose base orbifold is $D^2(p_\epsilon, q_\epsilon)$, and let Z be the connected submanifold of $M(\beta)$ which is the union of Z^+ and Z^- in $M(\beta)$ meeting along $\widehat{F}$. Then Z^ϵ is homeomorphic to $Q^\epsilon \times S^1$ where Q^ϵ is a twice punctured disk. Since $d = 1$, we have, as in the proof of Lemma 14.4, that Z is homeomorphic to the exterior of "the double Hopf link" in S^3 such that the torus $\widehat{F}$ becomes a torus in S^3 which bounds a trivial solid torus V^ϵ in S^3 on each side of $\widehat{F}$ and which separates the two pairs of parallel components of the double Hopf link and that the twice punctured disk Q^ϵ in Z^ϵ caps off in V^ϵ to a standard meridian disk of V^ϵ. Figure 14.3 shows Z^- cut open along Q^-. Note that $\widehat{X}^-$ is an annulus bundle over S^1 with A^- as a fibre. The part of A^- in Z^- is a twice punctured annulus. Figure 14.3 shows the part of A^- in the cut-open Z^- (the surface coloured green). Here we may assume that A^- is of the form shown in Figure 14.3 inside $Z = Z^+ \cup Z^-$ (i.e. after the specific choice of Q^- for Z^-) since there is only one twisted I-bundle over the Klein bottle up to homeomorphism.

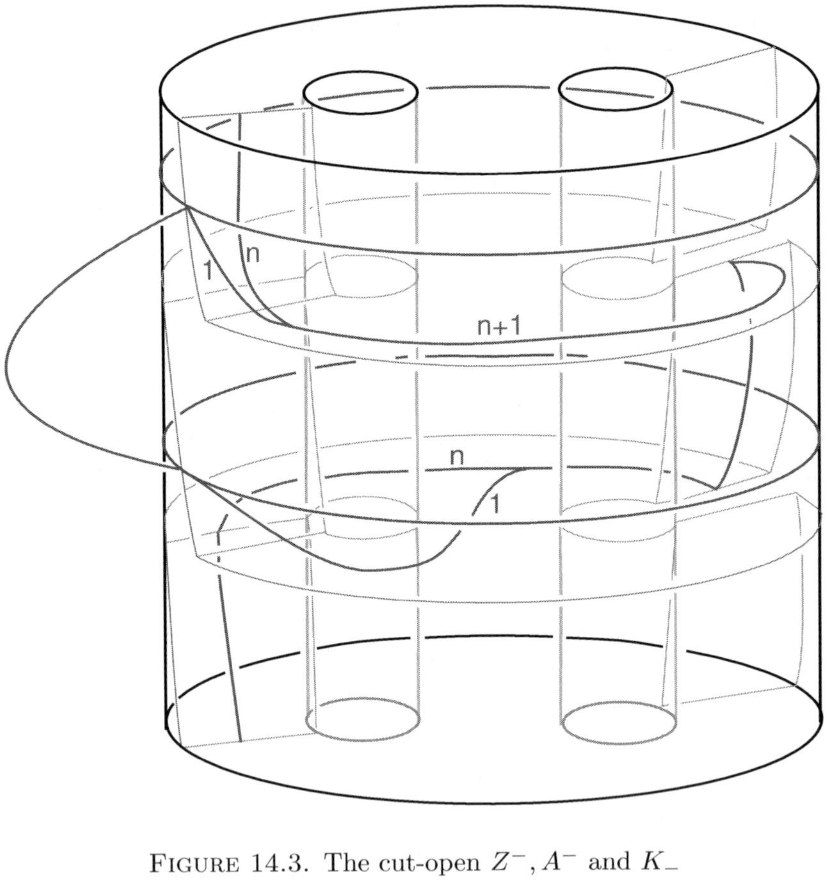

FIGURE 14.3. The cut-open Z^-, A^- and K_-

Also as in the proof of Lemma 14.4, we have that the annulus A^+ separates the two inner components of ∂Z^+, ∂A^+ is as shown in Figure 14.3 (the two horizontal circles coloured purple), and up to Dehn twists and isotopy, the arc k_+ in A^+ is transverse to the S^1 fibres of A^+ and is disjoint from a copy of Q^+ (shown in Figure 14.3 as the red coloured arc that lies outside Z^-). The arc k_- is an essential transverse arc in A^- which may wrap along the S^1 factor of A^- many times. In

Figure 14.3 k_- is illustrated as a branched curve in A^- (a branch with label n means n parallel arcs in A^-). To eliminate the wraps of k_- in A^-, we drill out two knots K_5, K_6 in Z^- which are chosen as follows: ∂A^- cuts the torus $\widehat{F}$ into two annuli, push the center circle of each of these two annuli slightly into the interior of Z^- but disjoint from A^-, the resulting knots are K_5 and K_6. Let Y^- be the exterior of K_5 and K_6 in Z^-, and let T_5 and T_6 be the torus boundary components of Y^- corresponding to K_5 and K_6. There is an obvious essential annulus A in Y^- connecting T_5 and T_6 which intersects A^- in a single circle which is essential in both A and A^-. Twisting along this annulus A (which does not change the homeomorphism type of Y^-, which fixes $\widehat{F}$ point-wise and which preserves A^- set-wise) together with some isotopy, we may simplify k_- in A^- so that it does not fully wrap around the S^1 factor of A^- and is disjoint from a copy of Q^-. Now let N be the exterior in S^3 of the 7-component link shown in part (1) of Figure 14.4 where the red coloured component K_0 is $K_\beta = k_+ \cup k_-$, the two green coloured components are K_5 and K_6 and the remaining four components form the double Hopf link. Then by the construction of N, M can be obtained by Dehn filling the six boundary components $T_1, \cdots, T_6$ of N, leaving T_0 unfilled.

Note that the twisted I-bundle over the Klein bottle $\widehat{X}^-$ can be recovered from Z^- by Dehn filling the two inner boundary tori T_3 and T_4 of Z^- both with slope $(-2, 1)$ since the twisting along the annulus A above fixes T_3 and T_4 point-wise and thus does not change filling slopes on these two tori.

Simplifying this link as illustrated in Figure 14.4 (which does not change the homeomorphism type of N), we get the link shown in Figure 14.4 part (2). By the Kirby-Rolfsen surgery calculus, the filling slope for the new T_3 and T_4 becomes $(-1, 1)$. This proves the first conclusion of the lemma.

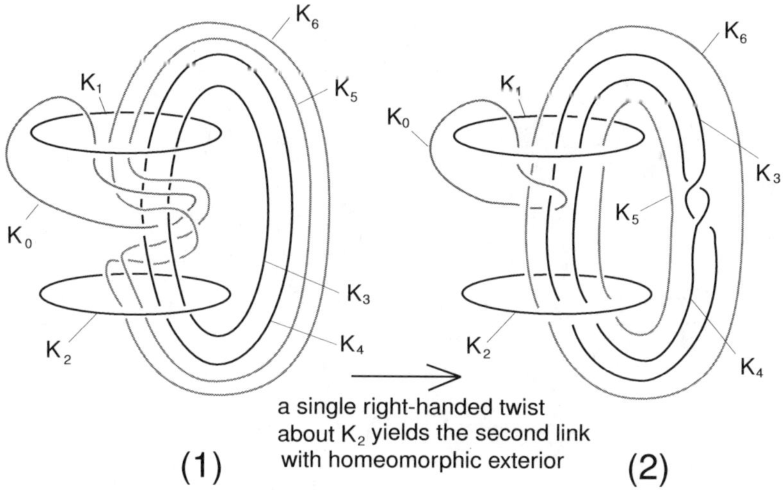

FIGURE 14.4. Part (1) shows the 7-component link in S^3 resulting from the argument of Lemma 14.6 whose exterior is homeomorphic to that of the link in S^3 shown in Part (2).

To prove the second conclusion of the lemma, let N be the exterior of the ordered 7-component link in S^3 shown in part (2) of Figure 14.4, which can be checked with SnapPy to be hyperbolic. Our manifold M is supposed to be a hyperbolic filling of N leaving T_0 unfilled, i.e.

$$M = N[(0,0),(p_1,q_1),(p_2,q_2),(-1,1),(-1,1),(p_5,q_5),(p_6,q_6)]$$

for some nonempty slopes $(p_1,q_1),(p_2,q_2),(p_5,q_5),(p_6,q_6)$. Let Y be the manifold

$$N[(0,0),(0,0),(0,0),(-1,1),(-1,1),(0,0),(0,0)].$$

There is an annulus E in Y joining T_5 and T_6 (since $\widehat{X}^-$ is an A^--bundle over S^1, K_5 and K_6 bound an annulus in $\widehat{X}^-$ disjoint from the annulus A^- which contains k_-). A regular neighbourhood Z_* of $E \cup T_5 \cup T_6$ in Y is homeomorphic to $P \times S^1$, where P is a twice punctured disk. Let $W = Y \setminus Z_*$, and let T be the torus $W \cap Z_*$. Our manifold M is obtained by Dehn filling Y along T_1, T_2, T_5 and T_6: $M = \widehat{W} \cup_T \widehat{Z}_*$, say, where $\widehat{W}$ is the result of Dehn filling W along T_1 and T_2 with slopes (p_1,q_1) and (p_2,q_2), and $\widehat{Z}_*$ is Z_* Dehn filled along T_5 and T_6 with slopes (p_5,q_5) and (p_6,q_6). Since M is hyperbolic, T compresses in either $\widehat{Z_*}$ or $\widehat{W}$. In the first case, $\widehat{Z}_*$ must be a solid torus, and in the second case, $\widehat{W} \cong M\#V$, where V is a solid torus and $V \cup \widehat{Z}_* \cong S^3$. In the first case, $\widehat{Z}_*$ can be obtained from Z_* by doing the trivial Dehn filling $(1,0)$ on T_6 and some filling on T_5, with slope (p_5',q_5'), say. In the second case, we can do a similar Dehn filling on Z_* to get a solid torus U such that $V \cup_T U \cong S^3$. Thus in both cases $M = N[(0,0),(p_1,q_1),(p_2,q_2),(-1,1),(-1,1),(p_5',q_5'),(1,0)]$. That is, if N_* is the exterior of the 6-component link in S^3 shown in part (1) of Figure 14.5, then $M = N_*[(0,0),(p_1,q_1),(p_2,q_2),(-1,1),(-1,1),(p_5',q_5')]$. Using the Kirby-Rolfsen surgery calculus we can eliminate the component K_4 and get the ordered 5-component link $\{J_0,J_1,J_2,J_3,J_4\}$ in S^3 shown in part (2) of Figure 14.5 so that $M = N_\#[(0,0),(m_1,n_1),(m_2,n_2),(m_3,n_3),(0,1)]$ for non-empty slopes $(m_1,n_1),(m_2,n_2),(m_3,n_3)$, where $N_\#$ is the exterior of the ordered 5-component link $\{J_0,J_1,J_2,J_3,J_4\}$ in S^3.

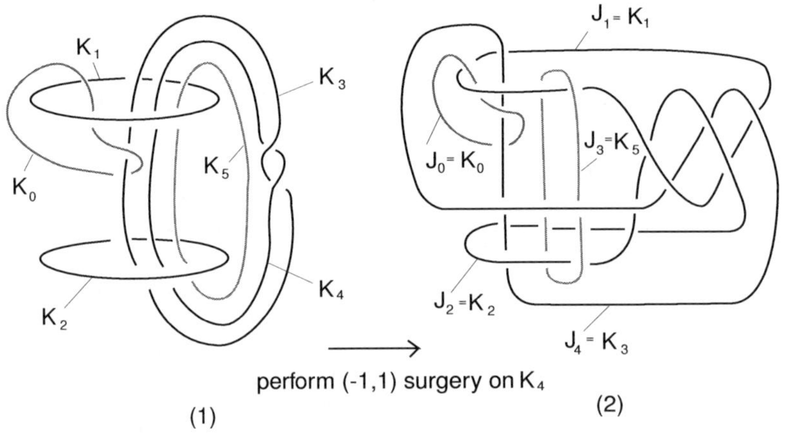

FIGURE 14.5. The ordered 5-component link $\{J_0,J_1,J_2,J_3,J_4\}$ in S^3

The five component link $\{J_0,J_1,J_2,J_3,J_4\}$ in S^3 is hyperbolic and we have reduced the proof of the lemma to the proof of the following claim.

CLAIM 14.7. If M can be obtained by Dehn filling $N_\#$ along $T_1,\dots,T_4$ with slope $(0,1)$ on T_4, then M is the figure 8 knot exterior.

It is enough to show that if a 4-filling

$$N_{\#}[(0,0),(m_1,n_1),(m_2,n_2),(m_3,n_3),(0,1)]$$

is a hyperbolic manifold which has two non-hyperbolic filling slopes on T_0 distance at least 6 apart, then it is either the figure eight knot exterior, or $Wh[(0,0),(-2,1)]$ or $Wh[(0,0),(5,1)]$, where Wh is the exterior of the Whitehead link in the 3-sphere,[1] because for each of $Wh[(0,0),(-2,1)]$ and $Wh[(0,0),(5,1)]$, if two non-hyperbolic filling slopes are distance at least 6 apart, they are both toroidal filling slopes.

To this end, we apply the Python codes *find_exceptional_fillings.py* and *search_-geometric_solutions* written by Martelli [**Ma**], used on SnapPy written by Culler, Dunfield, Goemer and Weeks [**CDGW**], to the 5-component link exterior $N_{\#}$. Note that when running *find_exceptional_fillings.py* on the 5-component link exterior $N_{\#}$ we only need to find those fillings of $N_{\#}$ which either have slope $(0,1)$ on T_4 or leave T_4 unfilled (this can be done by adding a finite set of well chosen 1-fillings into the exclude.py file in Martelli's Python package). We get two lists as the outputs from running the code *find_exceptional_fillings.py*: a list of candidate isolated non-hyperbolic fillings on $N_{\#}$ and a list of candidate hyperbolic fillings on $N_{\#}$, here an isolated non-hyperbolic filling means a non-hyperbolic filling whose proper sub-fillings are all hyperbolic fillings.

We first need to check if the fillings in the second list are indeed hyperbolic fillings, which is accomplished by running *search_-geometric_solutions* on it. It turns out the fillings in the second list are all 5-fillings (there are 862 of them) and they are all hyperbolic. (To reduce the running time, we first set the max degree in the code to 3, at which 809 fillings are confirmed to be hyperbolic, then set the max degree at 5, at which another twenty fillings are confirmed, then degree 7 confirming twenty-six more fillings. The remaining seven fillings are confirmed at degree 11.)

We now deal with the first list. We don't have to confirm if all the fillings listed are indeed non-hyperbolic. We treat them as possible isolated non-hyperbolic fillings. The good thing is that the list contains all real isolated non-hyperbolic fillings. The list contains eight 1-fillings, twenty-seven 2-fillings, sixty-three 3-fillings, one hundred and seventy 4-fillings and three hundred and one 5-fillings.

If a k-filling of $N_{\#}$, $1 \le k \le 3$, is an isolated non-hyperbolic filling but contains hyperbolic pieces, it might extend to infinitely many hyperbolic 4-fillings of $N_{\#}$ with T_0 unfilled, which could potentially contribute many new non-hyperbolic slopes on T_0 which have not occurred in the fillings of the list. Therefore such cases, if they exist, are potentially difficult to deal with. Fortunately such a case does not happen for our manifold $N_{\#}$. There are some k-fillings, $1 \le k \le 3$, which are isolated non-hyperbolic containing hyperbolic pieces. But after adding the slope $(0,1)$ on T_4 to such k-filling (if T_4 is unfilled there), it still contains a hyperbolic piece only when its slope on T_0 is non-empty and therefore this k-filling contributes at most one non-hyperbolic slope on T_0. Similarly we don't need to worry about isolated non-hyperbolic 4-fillings as each of them either contributes only one slope on T_0 or can not yield M.

Now we combine fillings in the first list (exclude those with empty slope on T_0) into the list of maximal subgroups so that in each subgroup any two fillings are compatible in the sense that their slopes on each of T_1, T_2, T_3, T_4 agree unless one

---

[1] For the rest of this chapter we've replaced the Whitehead link exterior Wh (cf. Figure 2.1) with its mirror image as a matter of convenience.

or both of them are empty. (The new list, obtained using a simple Python code, as well as the two lists mentioned earlier will be posted along with the paper on authors' web pages). Each such subgroup contains at least one 5-filling which by restriction gives a hyperbolic 4-filling leave T_0 unfilled (this hyperbolic 4-filling is unique for this subgroup). Slopes on T_0 from the elements in this subgroup contain all possible non-hyperbolic filling slopes for the hyperbolic 4-filling. So we just need to calculate the distance between such slopes on T_0 for each subgroup (which can be done using a simple Python code). It turns out that if the distance between two slopes on T_0 from a subgroup is at least 6, then one of the following events holds:

(1) The corresponding hyperbolic 4-filling for this subgroup is the figure eight knot exterior (occurs for 4 subgroups) or $Wh[(0,0),(5,1)]$ (occurs for 4 subgroups) or $Wh[(0,0),(-2,1)]$ (occurs for 8 subgroups), which can be verified by SnapPy.

(2) One of the two slopes is $(0,1)$. But this slope is not a non-hyperbolic slope of the corresponding 4-filling of the subgroup, which can be checked using *search_geometric_solutions*. (Such instance occurs for 11 subgroups). Such case happens because $[(0,1),(0,0),(0,0),(0,0),(0,0)]$ is a non-hyperbolic filling which contains a hyperbolic piece.

(3) The distance is 7 and is realized on the unique pair $[(-3,4),(2,1),(3,1),(-2,1),(0,1)]$ and $[(1,1),(2,1),(0,0),(0,0),(0,0)]$. But $[(1,1),(2,1),(3,1),(-2,1),(0,1)]$ is hyperbolic. This happens because $[(1,1),(2,1),(0,0),(0,0),(0,0)]$ is an isolated non-hyperbolic filling which contains hyperbolic piece.

The proof of the second conclusion of the lemma is now finished. $\diamondsuit$

14.3. The combinatorics of $\overline{\Gamma}_F$ and the figure eight knot exterior

We show how combinatorial conditions on $\overline{\Gamma}_F$ guarantee the existence of annuli as in the previous section.

PROPOSITION 14.8. *Suppose that $d \neq 0$, $\Delta(\alpha,\beta) \geq 6$, and f is a trigon face of $\overline{\Gamma}_F$ contained in X^ϵ with a maximal weight edge $\bar{e}$. Then there are annuli A^+ and A^- with the properties given in Lemma 14.4. Therefore M is the figure eight knot exterior.*

PROOF. Since f is a trigon, X^ϵ is not a twisted I-bundle and by Lemma 13.3, $\bar{e}$ is a positive edge.

Let e_1, e_2, e_3 be parallel adjacent edges of Γ_F corresponding to $\bar{e}$ where e_1 is an edge of f and e_2 is the edge lying between e_1 and e_3. Since $\bar{e}$ is a positive edge of weight 3 or more, $\widehat{\dot{\Sigma}_1^+}$ is a twisted I-bundle over a Möbuis band and $\widehat{\dot{\Sigma}_1^-}$ is either a twisted I-bundle over a Mobuis band or a twisted I-bundle over a Klein bottle.

After a homotopy of h we can assume that the face f of Γ_F is minimally positioned (cf. §13.3). Recall that A_ϵ is the frontier of $\widehat{\dot{\Sigma}_1^\epsilon}$ in $\widehat{X}^\epsilon$ with boundary components c_1 and c_2, $\widehat{P} = A_\epsilon \times I$ is a product region in $\widehat{\dot{\Sigma}_1^\epsilon}$ and $k_\epsilon = K_\beta \cap H^\epsilon$ is an essential arc in $A^\epsilon = A_\epsilon \times \{1/2\}$. By Corollary 13.8, e_1 can be homotoped in F with its two endpoints fixed so that it intersects each c_i at most once. Further, since the bigon face between e_2 and e_3 lies in X^ϵ, e_2 can be homotoped in F with its two endpoints fixed so that it is contained in the interior of $\dot{\Phi}_1^\epsilon$.

Recall that $\partial X^{-\epsilon} = F \cup B^{-\epsilon}$ and b_1, b_2 are components of $\partial F = \partial B^{-\epsilon}$.

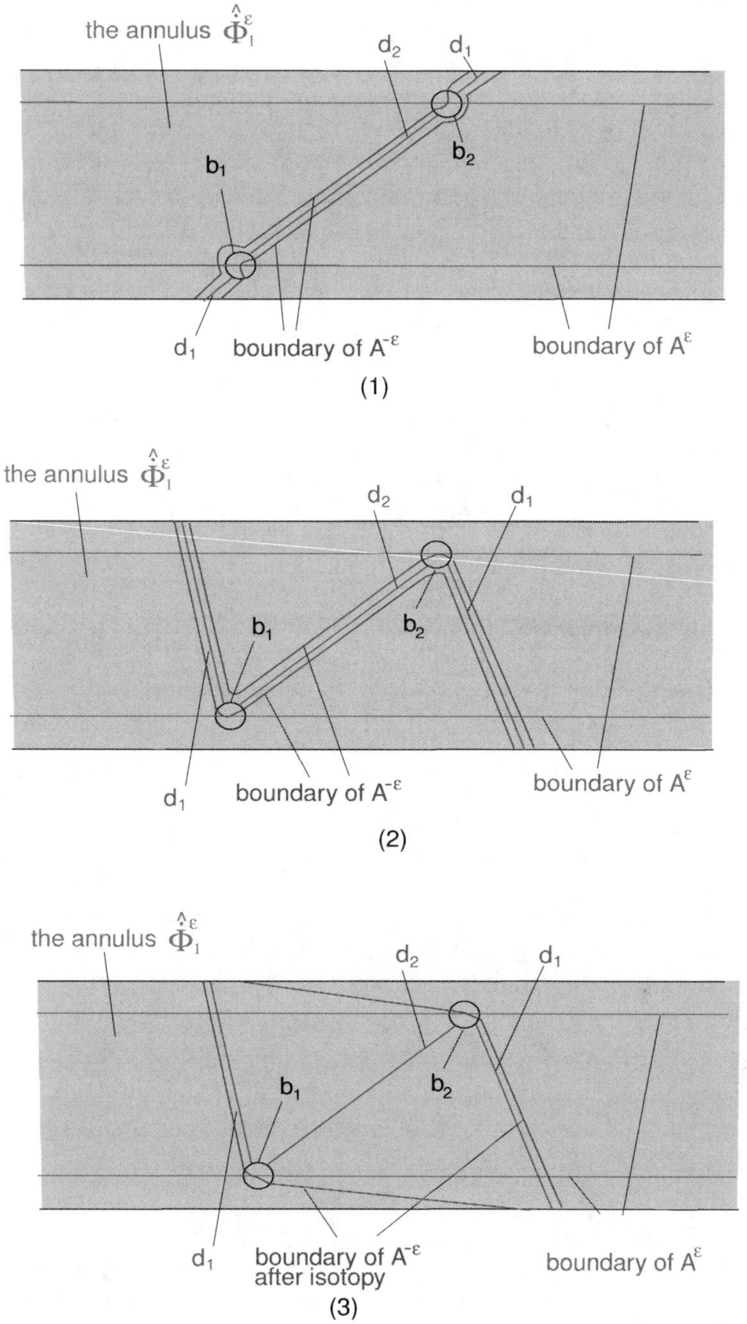

FIGURE 14.6. Illustration of the construction of the annuli A^{ϵ} and $A^{-\epsilon}$ with the required properties.

If we apply the loop theorem to the algebraic bigon R between e_1 and e_2 with respect to the algebraic intersection with b_1, we obtain an embedded disk D in $X^{-\epsilon}$

such that ∂D has non-zero algebraic intersection with b_1. In particular ∂D must intersect b_1. The loop theorem implies that we may assume that $\partial D \cap (B^{-\epsilon} \cup c_1 \cup c_2)$ is contained in $h(\partial R) \cap (B^{-\epsilon} \cup c_1 \cup c_2)$ (cf. [**He**, Theorem 4.10]). Lemma 11.1(1) shows that D cannot be a monogon, so it is an algebraic bigon. Each of the two components of $F \cap \partial D$ are arcs connecting b_1 and b_2 so that one of them, d_1 say, intersects each c_i at most once, and the other, d_2 say, is contained in the interior of $\dot{\Phi}_1^\epsilon$, and thus is disjoint from each c_i.

Note that D is part of an embedded Möbius band in $X^{-\epsilon}$ obtained by attaching a product region in the annulus $B^{-\epsilon}$ between the two corners of D (cf. the proof of Lemma 12.5). The boundary of this Möbius band has slope $\phi_{-\epsilon}$ and c_i has slope ϕ_ϵ. Our hypothesis that $d \neq 0$ implies that $\phi_+ \neq \phi_-$, so the boundary of the Möbius band intersects each c_i exactly once and the intersection is transverse. (So $d = 1$.) That is, the edge d_1 intersects each c_i transversely exactly once.

Recall $\hat{b}_1$ and $\hat{b}_2$ are the disks attached to F to form $\widehat{F}$. If we consider $\hat{b}_1$, $\hat{b}_2$ as points then $d_1 \cup \hat{b}_1 \cup d_2 \cup \hat{b}_2$ is a loop in $\widehat{F}$ of slope $\phi_{-\epsilon}$ which intersects c_i transversely exactly once. We may now isotope D in $X^{-\epsilon}$ so that its edges become transverse to the S^1-fibres of the annulus $\widehat{\dot{\Phi}_1^\epsilon}$. Figures 14.6(1) and (2) illustrate the possibilities for the edges d_1 and d_2 of D up to Dehn twists fixing their endpoints along a S^1-fibre of the annulus $\widehat{\dot{\Phi}_1^\epsilon}$. If the case of Figure 14.6(1) occurs, the resulting algebraic bigon D yields a corresponding embedded Möbius band B in $X^{-\epsilon}$ such that the frontier of a suitably chosen regular neighbourhood of B in $\widehat{X}^{-\epsilon}$ is an essential annulus $A^{-\epsilon}$ which contains $k_{-\epsilon}$ as an essential arc. Moreover each component of $\partial A^{-\epsilon}$ intersects each component of ∂A^ϵ exactly once. The boundary of $A^{-\epsilon}$ is illustrated in Figure 14.6(1).

If the case of Figure 14.6(2) occurs, the resulting algebraic bigon D also yields a corresponding embedded Möbius band B in $X^{-\epsilon}$ such that the frontier of a suitably chosen regular neighbourhood of B in $\widehat{X}^{-\epsilon}$ is an essential annulus $A^{-\epsilon}$ which contains $k_{-\epsilon}$ as an essential arc, as illustrated in Figure 14.6(2). But at this stage each component of $\partial A^{-\epsilon}$ intersects each component of ∂A^ϵ exactly twice, one transverse and one tangent. We may now isotope $A^{-\epsilon}$ in $\widehat{X^{-\epsilon}}$, fixing the arc $k_{-\epsilon} \subset A^{-\epsilon}$, so that at the end of the isotopy each component of $\partial A^{-\epsilon}$ intersects each component of ∂A^ϵ exactly once, and transversely, as indicated in Figure 14.6(3). $\diamondsuit$

PROPOSITION 14.9. *If $\Delta(\alpha, \beta) \geq 6$ and there is a quad face f of $\overline{\Gamma}_F$ with adjacent edges of maximal weight, then there are annuli A^+ and A^- with the properties given in Lemma 14.4. Therefore M is the figure eight knot exterior.*

PROOF. Suppose that f lies in X^ϵ and note that X^ϵ cannot be a twisted I-bundle as in this case the only faces of Γ_F lying in X^- are bigons. We may assume that f is minimally positioned (cf. §13.3).

By Corollary 13.9, f has an edge $\bar{e}$ of maximal weight which intersects each c_i at most once. Let e_1, e_2, e_3 be the parallel adjacent edges in the family of edges represented by $\bar{e}$ with e_1 being the edge of f and e_2 being the one between e_1 and e_3. Note that we may assume that e_2 is contained in the interior of $\dot{\Phi}_1^\epsilon$ and thus is disjoint from each c_i.

Let R be the bigon face between e_1 and e_2. Note that ∂R is an essential loop in $\partial X^{-\epsilon}$ since each of the edges e_1 and e_2 is an essential arc in F, so we can apply the loop theorem to the singular disk R which is the bigon face between e_1 and e_2. We obtain a properly embedded disk $(D, \partial D)$ in $(X^{-\epsilon}, \partial X^{-\epsilon})$ such that ∂D

is an essential loop in $\partial X^{-\epsilon}$ which intersects each c_i transversely and in at most one point. Since F is incompressible and $X^{-\epsilon}$ does not contain a monogon, D is a bigon.

If D is an algebraic 0-gon, its edges are disjoint $\widehat{F}$-essential loops and therefore their algebraic intersection with each c_i is even, contrary to what we have deduced. Thus it is an algebraic bigon. Now proceed as in the proof of Proposition 14.8 to see that the existence of such an embedded disk D implies that there are annuli A^+ and A^- with the properties given in Lemma 14.4. Therefore M is the figure eight knot exterior. ♢

PROPOSITION 14.10. *Suppose that $\Delta(\alpha, \beta) = 6$ and $d = 1$. If $\overline{\Gamma}_F$ has two trigon faces f_1 and f_2 sharing a common positive edge $\bar{e}$ of weight 2, then M is the figure eight knot exterior.*

PROOF. Let e_1 and e_2 be the two parallel edges of Γ_F represented by $\bar{e}$. Note that f_1 and f_2 lie on the same side of F, say X^ϵ, and as they are trigons, X^ϵ is not a twisted I-bundle. We may assume that each of e_1 and e_2 intersects each c_i exactly once by Corollaries 13.6(2) and 13.8.

The bigon R containing e_1 and e_2 in its boundary is an S-bigon. That is, e_1 and e_2 form an S-cycle. Applying the loop theorem to the singular disk R with respect to (mod 2) intersection with b_1 yields an embedded S-bigon D in $X^{-\epsilon}$ the union of whose two edges intersects each c_i at most twice. As in the proof of Proposition 14.8, the bigon yields a Möbius band in $X^{-\epsilon}$ whose boundary has slope $\phi_{-\epsilon}$ and intersects each c_i at most twice. Since $d = 1$, we see that the boundary of the Möbius band intersects each c_i exactly once and the intersection is transverse. As in the proof of Proposition 14.8, such a Möbius band will yield an embedded essential annulus $A^{-\epsilon}$ which together with A^ϵ, chosen as in the proof of Proposition 14.8, satisfy the properties of Lemma 14.4 and thus M is the figure eight knot exterior. ♢

PROPOSITION 14.11. *Suppose that $\Delta(\alpha, \beta) = 6$, $d = 1$ and X^- is a twisted I-bundle. If $\overline{\Gamma}_F$ has two trigon faces f_1 and f_2 sharing a common edge $\bar{e}$, then M is the figure eight knot exterior.*

PROOF. Since X^- is a twisted I-bundle, both f_1 and f_2 lie in X^+. By Proposition 14.8, we may assume that $\bar{e}$ has weight 2.

Let e_1 and e_2 be the two parallel edges of Γ_F represented by $\bar{e}$. By Corollary 13.8, we may assume that each of e_1 and e_2 intersects each c_i at most once. Now applying the loop theorem to the bigon face R of Γ_F containing e_1 and e_2 in its boundary to obtain an embedded bigon D in X^-. The boundary of D intersects each c_i transversely and in at most two points.

CASE 1. D is an algebraic 0-gon.

In this case, the edges of D are disjoint $\widehat{F}$-essential loops and therefore their union intersects each c_i transversely in at most two points. Since these two loops are homologous, either both are disjoint from the c_i or both intersect each c_i exactly once.

As in the third paragraph of the proof of Lemma 12.6, the edges of D form a pair of $\widehat{F}$-essential loops which are the boundary of an essential annulus A^- in X^-. Thus they have slope ϕ_- or ϕ'_-. Since $d = 1$ and $d' \neq 0$ (cf. Proposition 13.2), each edge of D, which is an $\widehat{F}$-essential loop in F, intersects each c_i exactly once and the intersection is transverse. In particular, in the case that the edges of D have slope ϕ'_-, we have $d' = 1$. The reader will verify that A^- and A^+ can be positioned to have the properties listed in the statement of Lemma 14.4 when ∂A^- has slope ϕ_-, or those listed in Lemma 14.6 when ∂A^- has slope ϕ'_-. Hence M is the figure eight knot exterior.

CASE 2. D is an algebraic bigon.

The argument is similar to that given in the proof of Proposition 14.8. ♢

The proof of Proposition 14.11 actually shows the following result.

COROLLARY 14.12. *Suppose that* $\Delta(\alpha, \beta) = 6$, $d = 1$, *and* X^- *is a twisted* I*-bundle. If* Γ_F *has a bigon face in* X^- *the union of whose two edges are incident to at most two red tiles, then* M *is the figure eight knot exterior.* ♢

PROPOSITION 14.13. *If* $\Delta(\alpha, \beta) = 6$, $d = 1$, X^- *is a twisted* I*-bundle, and there is a* 5*-gon face* f *of* $\overline{\Gamma}_F$ *with adjacent edges of weight* 4*, then* M *is the figure eight knot exterior.*

PROOF. Note that f lies in X^+, so we may assume that f is minimally positioned (cf. §13.3). Let e_1 and e_2 be adjacent edges of f in Γ_F which have maximal weight. Suppose that we can show that there are at most two red tiles incident to one of these edges, e_1 say. Since $q_+ \geq 3$, no bigon face of Γ_F lying in X^+ contains a red tile, so we can apply Corollary 14.12 to the bigon face of Γ_F incident to e_1 to complete the proof.

Suppose that there are three red tiles incident to each of e_1 and e_2. Since $q_+ \geq 3$, there are no red tile bigons connecting e_1 to either of its adjacent edges in ∂f, and a similar statement holds for e_2. It follows that for $i = 1, 2$, a red tile incident to e_i contains a cross-arc in its boundary running between e_i and the edge e' of f which is adjacent to neither e_1 nor e_2 (cf. Figure 14.7). By Lemma 13.10, each cross-arc connecting e_1 and e' has the same labels at its endpoints, and the same holds for the cross-arcs connecting e_2 and e'. But then the restriction of h to each of these arcs is homotopic (rel endpoints) to a path in F and so we can construct a monogon in X^+ (cf. Figure 14.7); the sub-disk of f between the cross-arcs a and a' with a corner at v is a monogon), which gives a contradiction. ♢

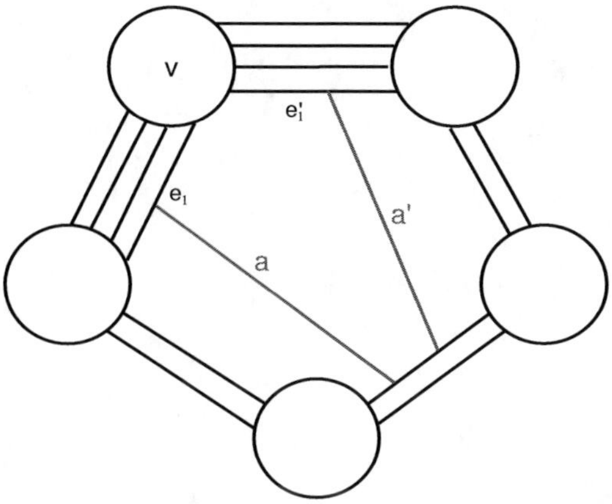

FIGURE 14.7. The sub-disk between the cross-arcs a and a' with a corner at v is a monogon.

CHAPTER 15

Completion of the proof of Theorem 1.2 when $\Delta(\alpha, \beta) \geq 7$

To complete of the proof of Theorem 1.2 when $\Delta(\alpha, \beta) \geq 7$, we must prove the following proposition.

PROPOSITION 15.1. *Suppose that Assumptions* 4.3 *and* 9.2 *hold, F is separating but not a semi-fibre, and $t_1^+ = t_1^- = 0$. If $\Delta(\alpha, \beta) \geq 7$, then M is the figure eight knot exterior.*

PROOF. Since the edges of $\overline{\Gamma}_F$ have weight at most 4 (Lemma 13.3), for each vertex v of Γ_F we have an inequality

$$14 \leq 2\Delta(\alpha, \beta) = \text{valency}_{\Gamma_F}(v) \leq 4 \text{ valency}_{\overline{\Gamma}_F}(v),$$

which shows that the valency of v in $\overline{\Gamma}_F$ is at least 4.

First suppose that there is a vertex v of Γ_F such that $\mu(v) > 2\Delta(\alpha, \beta) - 4$. Propoition 12.2 of [**BGZ2**] then shows that the valency of v is at most 5 and

$$\varphi_3(v) \geq \begin{cases} 1 & \text{if valency}(v) = 4 \\ 4 & \text{if valency}(v) = 5 \end{cases}$$

Consequently, the condition that $\Delta(\alpha, \beta) \geq 7$ implies that there is a trigon face of $\overline{\Gamma}_F$ incident to v with an edge of maximal weight and therefore M is the figure eight knot exterior by Proposition 14.8.

Thus we may assume that $\mu(v) \leq 2\Delta(\alpha, \beta) - 4$ for each v. By Corollary 12.4 of [**BGZ2**], $\mu(v) = 2\Delta(\alpha, \beta) - 4$ for each v and then Proposition 12.5 of [**BGZ2**] implies that if v is a vertex of $\overline{\Gamma}_F$, then either

- valency$(v) = 4 = \varphi_4(v)$, or
- valency$(v) = 5$, $\varphi_3(v) = 3$, and $\varphi_4(v) = 2$, or
- valency$(v) = 6 = \varphi_3(v)$.

Hence if $\overline{\Gamma}_F$ has a vertex of valency 5 or 6, our assumption that $\Delta(\alpha, \beta) \geq 7$ implies that it is incident to a trigon face with an edge of maximal weight, so we are done by Proposition 14.8.

Assume then that valency$(v) = 4 = \varphi_4(v)$ for each vertex v. Then $\overline{\Gamma}_F$ is rectangular and as the weight of the edges is 4 or less, $\Delta(\alpha, \beta)$ is either 7 or 8. It follows that each vertex v of $\overline{\Gamma}_F$ is incident to at least three edges of weight 4, so there is a quad face of $\overline{\Gamma}_F$ with adjacent edges of maximal weight. Proposition 14.9 then shows that M is the figure eight knot exterior. $\diamondsuit$

CHAPTER 16

Completion of the proof of Theorem 1.2 when X^- is not a twisted I-bundle

In this chapter we complete of the proof of Theorem 1.2 when X^- is not a twisted I-bundle.

PROPOSITION 16.1. *Suppose that Assumptions 4.3 and 9.2 hold, F is separating but not a semi-fibre, and $t_1^+ = t_1^- = 0$. If X^- is not a twisted I-bundle, then $\Delta(\alpha, \beta) \leq 5$.*

PROOF. Since X^- is not a twisted I-bundle, M is not the figure eight knot exterior.

Suppose that $\Delta(\alpha, \beta) \geq 6$ and note that $\Delta(\alpha, \beta) = 6$ by Proposition 15.1. Proposition 13.1 implies that $d > 0$. The weight of each edge in $\overline{\Gamma}_F$ is at most 3 by Lemma 13.3 and therefore $\overline{\Gamma}_F$ has no valency 3 vertices and each edge incident to a valency 4 vertex is of maximal weight.

If there is a vertex v of Γ_F such that $\mu(v) > 2\Delta(\alpha, \beta) - 4$, Proposition 12.2(1) of [**BGZ2**] implies that $\overline{\Gamma}_F$ has a trigon face with an edge of weight 3. But then M would be the figure eight knot exterior by Proposition 14.8, a contradiction.

Corollary 12.4 of [**BGZ2**] now shows that $\mu(v) = 2\Delta(\alpha, \beta) - 4$ for each vertex v and by Proposition 12.5 of that paper, for each vertex v of $\overline{\Gamma}_F$ either

- valency$(v) = 4 = \varphi_4(v)$, or
- valency$(v) = 5$, $\varphi_3(v) = 3$, and $\varphi_4(v) = 2$, or
- valency$(v) = 6 = \varphi_3(v)$.

Applying Propositions 14.8 and 14.9 we see that $\overline{\Gamma}_F$ has no vertices of valency 4 or 5. It follows that $\overline{\Gamma}_F$ is hexagonal. Each of its edges has weight 2 by Proposition 14.8 while Proposition 11.2 of [**BGZ2**] shows that some edge $\bar{e}$ of $\overline{\Gamma}_F$ is negative. Then $d = 1$ by Lemma 13.3(1)(c) and as every trigon face of $\overline{\Gamma}_F$ has a positive edge we can apply Proposition 14.10 to see that M is the figure eight knot exterior. This final contradiction completes the proof. ◊

CHAPTER 17

Completion of the proof of Theorem 1.2 when $\Delta(\alpha, \beta) = 6$ and $d = 1$

We suppose that F is separating, $\Delta(\alpha, \beta) = 6$, $t_1^+ = t_1^- = 0$, and $d = 1$ in this chapter. By Proposition 16.1 we can assume that X^- is a twisted I-bundle and therefore each edge of $\overline{\Gamma}_F$ has weight 2 or 4 by Lemma 13.3. We prove,

PROPOSITION 17.1. *Suppose that Assumptions 4.3 and 9.2 hold, F is separating but not a semi-fibre, and $t_1^+ = t_1^- = 0$. If $\Delta(\alpha, \beta) = 6$ and $d = 1$, then M is the figure eight knot exterior.*

The proof of this proposition follows from Lemmas 17.2 and 17.3 below. In order to set these lemmas up, recall that $0 \le \chi(Y) = \sum_v \chi(v)$ where the sum is over the vertices of $\overline{\Gamma}_F$ and

$$\chi(v) = 1 - \frac{\text{valency}(v)}{2} + \sum_{v \in f} \frac{\chi(f)}{|\partial f|}$$

LEMMA 17.2. *Suppose that X^- is a twisted I-bundle, $\Delta(\alpha, \beta) = 6$, and $d = 1$. If $\overline{\Gamma}_F$ has a vertex v with $\chi(v) > 0$, then M is the figure eight knot exterior.*

PROOF. Consider a vertex v of $\overline{\Gamma}_F$ such that $\chi(v) > 0$. Then

$$0 < \chi(v) = 1 - \frac{\text{valency}(v)}{2} + \sum_{v \in f} \frac{\chi(f)}{|\partial f|} \le 1 - \frac{\text{valency}(v)}{2} + \frac{\text{valency}(v)}{3} < 1 - \frac{\text{valency}(v)}{6}$$

so that valency$(v) \le 5$. On the other hand, the assumption that $\Delta(\alpha, \beta) = 6$ implies that the vertices of $\overline{\Gamma}_F$ have valency 3 or more. Thus the valency of v is either $3, 4$, or 5.

CASE 1. v has valency 5. The reader will verify that $\chi(v) > 0$ is equivalent to requiring that $\varphi_3(v) \ge 4$. Then each edge incident to v is incident to a trigon face of $\overline{\Gamma}_F$. On the other hand, as $\Delta(\alpha, \beta) = 6$ and v has valency 5, there is at least one edge incident to v having weight 4. Proposition 14.8 then shows that M is the figure eight knot exterior.

CASE 2. v has valency 4. Since $\Delta(\alpha, \beta) = 6$, two edges incident to v have weight 2 and two have weight 4. The condition that $\chi(v) > 0$ implies that $\varphi_3(v) \ge 1$. By Proposition 14.8, we may assume that $\varphi_3(v) = 1$ and the weights of the edges incident to v are $2, 2, 4, 4$ in cyclic order. The condition that $\chi(v) > 0$ then implies that either $\varphi_4(v) = 3$ or $\varphi_4(v) = 2$ and $\varphi_5(v) = 1$. In the former case there is a quad face incident to v which has adjacent edges of maximal weight, and so Proposition 14.9 shows that M is the figure eight knot exterior. In the latter case there is either a quad face or a 5-gon face incident to v with adjacent edges of weight 4. We may now apply Propositions 14.9 and 14.13 to see that M is the figure eight knot exterior.

CASE 3. v has valency 3. Since $\Delta(\alpha, \beta) = 6$, each of the edges incident to v has weight 4. The condition that $\chi(v) > 0$ shows that at least one of the faces incident to v is a 3-gon or 4-gon or 5-gon. So we may apply Propositions 14.8, 14.9 and 14.13 respectively to see that M is the figure eight knot exterior. ♢

LEMMA 17.3. *Suppose that X^- is a twisted I-bundle, $\Delta(\alpha, \beta) = 6$, and $d = 1$. If $\chi(v) \leq 0$ for each vertex v of $\overline{\Gamma}_F$, then M is the figure eight knot exterior.*

PROOF. Since $0 \leq \chi(Y) = \sum_v \chi(v)$, we have $\chi(v) = 0$ for each vertex v. One can easily verify that for any vertex v with $\chi(v) = 0$, one of the following four cases holds:

- valency$(v) = 3$ and the number of edges of the three faces incident to v are $(4, 6, 12), (4, 8, 8)$, $(5, 5, 10)$, or $(6, 6, 6)$;
- valency$(v) = 4 = \varphi_4(v)$;
- valency$(v) = 5$, $\varphi_3(v) = 3$, and $\varphi_4(v) = 2$;
- valency$(v) = 6 = \varphi_3(v)$.

By Proposition 14.11, we may assume that $\overline{\Gamma}_F$ has no vertices of valency 5 or 6.

Suppose that $\overline{\Gamma}_F$ has a vertex v of valency 3. Then each of the edges incident to v have weight 4, so by Propositions 14.8, 14.9 and 14.13 we may assume that $\varphi_6(v) = 3$. It follows that v cannot be connected to a valency 4 vertex by an edge of $\overline{\Gamma}_F$. In particular, the vertices of $\overline{\Gamma}_F$ which cobound an edge with v have valency 3. It follows that if f is a 6-gon faces incident to v, each of its edges has weight 4.

Since $q_+ \geq 3$, no bigon face of Γ_F lying in X^+ contains a red tile and there are no red tile bigons connecting adjacent edges of f. It follows that if an edge e of f has at least two red tiles incident to it, each of these tiles has a cross-arc in its boundary connecting e to another edge of f. But then it's easy to see that some edge of f is incident to at most one red tile. Corollary 14.12 then shows that M is the figure eight knot exterior.

Finally suppose that $\overline{\Gamma}_F$ has no valency 3 vertices. In this case, $\overline{\Gamma}_F$ is rectangular ([**BGZ2**, Proposition 11.5]) and the weights of edges at any of its vertices alternate 2, 4, 2, 4 (Proposition 14.9). By Corollary 14.12 we may assume that every edge of weight 4 intersects cross-arcs and thus there are no cross-arcs incident to a weight 2 edge. It follows that the two edges of a bigon face corresponding to a weight 2 edge is incident to at most two red tiles. Corollary 14.12 then implies that M is the figure 8 exterior. ♢

CHAPTER 18

The case that F separates but not a semi-fibre, $t_1^+ = t_1^- = 0$, $d \neq 1$, and $M(\alpha)$ is very small

In this chapter we suppose that F is separating, though not a semi-fibre, $t_1^+ = t_1^- = 0$, $d \neq 1$, and $M(\alpha)$ is a small Seifert manifold which is very small. We use character variety methods to prove the following proposition.

PROPOSITION 18.1. *Suppose that Assumptions* 4.3 *and* 9.2 *hold, F is separating but not a semi-fibre, $t_1^+ = t_1^- = 0$, and $d \neq 1$. If $M(\alpha)$ is a very small Seifert manifold, then*

$$\Delta(\alpha,\beta) \leq \begin{cases} 1 & \text{if } d = 0 \text{ or } M(\alpha) \text{ is of } C\text{-type} \\ 2 & \text{if } \alpha \text{ is of } D\text{-type} \\ 3 & \text{if } \alpha \text{ is of } T\text{-type, } O\text{-type or } I\text{-type,} \\ & \text{or } (a,b,c) = (2,3,6), (2,4,4) \text{ or } (3,3,3) \end{cases}$$

The case that $d = 0$ is handled in Proposition 13.1, so we suppose that $d \geq 2$ below.

Recall that $\widehat{X}^\epsilon$ has base orbifold $D^2(p_\epsilon, q_\epsilon)$ where $2 \leq p_\epsilon \leq q_\epsilon$. Write $\pi_1(D^2(p_\epsilon, q_\epsilon)) \cong \mathbb{Z}/p_\epsilon * \mathbb{Z}/q_\epsilon$ and choose generators a_ϵ of $\mathbb{Z}/p_\epsilon$ and b_ϵ of $\mathbb{Z}/q_\epsilon$ for which $a_\epsilon b_\epsilon$ generates $\pi_1(\partial D^2(p_\epsilon, q_\epsilon))$. There is an epimorphism

$$\pi_1(M(\beta)) \xrightarrow{\varphi} \Delta(p_+, q_+, d) *_\psi \Delta(p_-, q_-, d)$$

where φ is the quotient by $\langle\langle \phi_+, \phi_- \rangle\rangle_{\pi_1(M(\beta))}$ and $\psi : \varphi(\pi_1(\widehat{F})) \to \varphi(\pi_1(\widehat{F}))$ is the isomorphism determined by the gluing map $\partial \widehat{X}^+ \to \partial \widehat{X}^-$. The reader will verify that for either ϵ, $\varphi(\pi_1(\widehat{F}))$ is the copy $\mathbb{Z}/d$ contained in $\Delta(p_\epsilon, q_\epsilon, d)$ corresponding to $(a_\epsilon b_\epsilon)^d$.

If $\beta^* \in \pi_1(\partial M)$ is a dual class to β, then $\varphi(\beta^*) = a_+(a_+b_+)^k a_-$ for some integer k.

LEMMA 18.2. *Suppose that $t_1^+ = t_1^- = 0$ and $d \geq 2$. If $d' > 1$ is a divisor of d, there is a non-trivial curve $X_{d'} \subset X_{PSL_2}(M(\beta)) \subset X_{PSL_2}(M)$ such that*

(1) *$\tilde{f}_{\beta^*}$ has a pole at each ideal point of $X_{d'}$;*
(2) (a) *$X_{d'}$ has one ideal point if $(p_-, q_-, d') = (2,2,2)$, and therefore $s_{X_{d'}} \geq 1$;*
 (b) *$X_{d'}$ has two ideal points if $(p_-, q_-, d') \neq (2,2,2)$, and therefore $s_{X_{d'}} \geq 2$.*
(3) *$X_{d'}$ is not strictly non-trivial if and only if $(p_+, q_+, d') = (2,4,4)$ and (p_-, q_-, d') is either $(2,2,4)$ or $(2,4,4)$.*

PROOF. If $d' > 1$ is a positive integer dividing d we can use the surjective composition of homomorphisms

$$\pi_1(M(\beta)) \xrightarrow{\varphi} \Delta(p_+, q_+, d) *_{\mathbb{Z}/d} \Delta(p_-, q_-, d) \to \Delta(p_+, q_+, d') *_{\mathbb{Z}/d'} \Delta(p_-, q_-, d')$$

to construct a curve $X_{d'} \subset X_{PSL_2}(M(\beta)) \subset X_{PSL_2}(M)$. Lemma 6.3(3) shows that assertion (1) of the lemma holds and further, that $X_{d'}$ has two ideal points unless $(p_\epsilon, q_\epsilon, d) = (2,2,2)$ for some ϵ, and otherwise one. Since $(p_+, q_+) \neq (2,2)$, assertion (2) of the lemma holds. Lemma 6.3(2) shows that $X_{d'}$ is not strictly non-trivial if and only if $(p_\epsilon, q_\epsilon, d') \in \{(2,2,d'), (2,4,4)\}$ for both ϵ. Since $(p_+, q_+) \neq (2,2)$, the latter condition is equivalent to $(p_+, q_+, d') = (2,4,4)$ and $(p_-, q_-, d') \in \{(2,2,4), (2,4,4)\}$. Thus assertion (3) of the lemma holds. ◇

PROOF OF PROPOSITION 18.1. As mentioned above, the case $d = 0$ is handled by Proposition 13.1. Suppose then that $d \geq 2$.

If $d' > 1$ is a divisor of d and $X_{d'} \subset X_{PSL_2}(M)$ is a non-trivial curve obtained as in Lemma 18.2, then

$$\Delta(\alpha, \beta) s_{X_{d'}} = \|\alpha\|_{X_{d'}}$$

by (5.2) and further, $s_{X_{d'}} \neq 0$. Hence if $\pi_1(M(\alpha))$ is cyclic, then $\|\alpha\|_{X_{d'}} = s_{X_{d'}}$ by [**BCSZ2**, Proposition 8.1], so $\Delta(\alpha, \beta) = 1$.

Next suppose that $M(\alpha)$ is a prism manifold. That is, α is of D-type. If $X_{d'}$ is strictly non-trivial, then $\Delta(\alpha, \beta) s_{X_{d'}} = \|\alpha\|_X \leq 2 s_{X_{d'}}$ ([**BCSZ2**, Proposition 8.1]), so $\Delta(\alpha, \beta) \leq 2$. On the other hand, if $X_{d'}$ is not strictly non-trivial, then $(p_+, q_+, d') = (2,4,4)$ and (p_-, q_-, d') is $(2,2,4)$ or $(2,4,4)$ by Lemma 18.2(3). In either case, X_2 is a strictly non-trivial curve, so $\Delta(\alpha, \beta) \leq 2$ as above.

Finally suppose that $M(\alpha)$ is neither C-type nor D-type and recall from the proof of Proposition 10.7 that if $\rho : \pi_1(M(\alpha)) \to PSL_2(\mathbb{C})$ is an irreducible representation, then

$$\text{image}(\rho) \cong \begin{cases} T_{12} & \text{if } \alpha \text{ is of } T\text{-type or } (a,b,c) = (3,3,3) \\ D_3 \text{ or } O_{24} & \text{if } \alpha \text{ is of } O\text{-type} \\ I_{60} & \text{if } \alpha \text{ is of } I\text{-type} \\ D_3 \text{ or } T_{12} & \text{if } (a,b,c) = (2,3,6) \\ D_2 \text{ or } D_4 & \text{if } (a,b,c) = (2,4,4) \end{cases}$$

Further, in each of these five possibilities at most two such representations have isomorphic images and if two, either α has I-type and the image is I_{60} or $(a,b,c) = (2,4,4)$ and the image is D_4. Note, in particular, that the image of ρ is finite with elements of order at most 5.

Suppose that $(p_-, q_-, d') \neq (2,2,2)$, so that $s_{X_{d'}} \geq 2$ and no representation with character lying on $X_{d'}$ has image isomorphic to D_2. If $X_{d'}$ is strictly non-trivial, we can apply Proposition 5.2(2) to see that $\Delta(\alpha, \beta) \leq 2$ when α is of T-type, O-type or (a,b,c) is either $(2,3,6), (2,4,4)$, or $(3,3,3)$ and that $\Delta(\alpha, \beta) \leq 3$ when it is of I-type.

If $(p_-, q_-, d') \neq (2,2,2)$ and $X_{d'}$ is not strictly non-trivial, then as in the prism manifold case we know that $(p_+, q_+, d') = (2,4,4)$ and (p_-, q_-, d') is either $(2,2,4)$ or $(2,4,4)$. In either case, X_2 is a strictly non-trivial curve with $s_{X_2} \geq 2$ and as in the previous paragraph, $\Delta(\alpha, \beta) \leq 2$ when α is of T-type, O-type or (a,b,c) is either $(2,3,6), (2,4,4)$, or $(3,3,3)$, and that $\Delta(\alpha, \beta) \leq 3$ when it is of I-type.

Next suppose that $(p_-, q_-, d') = (2,2,2)$. In this case $X_{d'} = X_2$ is strictly non-trivial and the image of each representation whose character lies on X_2 contains a copy of $\Delta(2,2,2)$ as a proper subgroup, which excludes the possibility that the image is D_2 or D_3. It then follows from Proposition 5.2(2) that $\Delta(\alpha,\beta) \leq 3$ when α is of T-type or O-type or (a,b,c) is either $(2,3,6),(2,4,4)$ or $(3,3,3)$.

Finally suppose that α has I-type. Proposition 5.2 shows that to obtain the inequality $\Delta(\alpha,\beta) \leq 3$, it suffices to prove $|J_{X_2}(\alpha)| \leq 1$.

Suppose then that $|J_{X_2}(\alpha)| = 2$. There cannot be another non-trivial curve $X_* \subset X_{PSL_2}(M(\beta)) \subset X_{PSL_2}(M)$ as otherwise $|J_{X_*}(\alpha)| = \emptyset$, so $\Delta(\alpha,\beta) = 1$ (Proposition 5.2). In particular, since killing $\pi_1(\widehat{F})$ in $\pi_1(M(\beta))$ yields $\mathbb{Z}/\gcd(p_+,q_+) * \mathbb{Z}/2$, we must have $\gcd(p_+,q_+) = 1$. Also, the discussion in §6.1 shows that $(p_+,q_+) = (2,3)$, so $\widehat{X}^+$ is a trefoil exterior and $(p_+,q_+,d) = (2,3,2)$. We claim that there are no I-type filling slopes in this situation.

The assumption that $(p_-,q_-,d') = (2,2,2)$ implies that X^- is a twisted I-bundle ([**BGZ2**, Proposition 7.5]) and therefore X^- is a twisted I-bundle over the Klein bottle. Then $H_1(\widehat{X}^-) \cong \mathbb{Z} \oplus \mathbb{Z}/2$ where the $\mathbb{Z}$ factor is generated by a class ξ such that $2\xi = [\phi_-]$ and the $\mathbb{Z}/2$ factor is generated by $[\phi'_-]$.

Now if $[\phi_+] = m[\phi_-] + n[\phi'_-]$, then $|n| = \Delta(\phi_+,\phi_-) = d = 2$, so $[\phi_+] = 2m\xi \pm 2[\phi'_-] \in H_1(\widehat{X}^-)$. Hence $[\phi_+]$ maps to zero in $H_1(\widehat{X}^-;\mathbb{Z}/2)$. On the other hand, as the fibre class in a trefoil knot exterior, $[\phi_+]$ maps to zero in $H_1(\widehat{X}^+;\mathbb{Z}/2)$. Consideration of the Mayer-Vietoris sequence for $M(\beta) = \widehat{X}^+ \cup \widehat{X}^-$ then shows that the first (mod 2) Betti number of $M(\beta)$ is 2. Hence that of M is at least 2, and so that of $M(\alpha)$ is at least 1. In particular, α cannot be an I-type filling slope (cf. [**BZ1**, page 117]), contrary to our assumptions.

We conclude that $|J_{X_2}(\alpha)| \leq 1$ and therefore $\Delta(\alpha,\beta) \leq 3$. $\diamondsuit$

CHAPTER 19

The case that F separates but is not a semi-fibre, $t_1^+ = t_1^- = 0$, $d > 1$, and $M(\alpha)$ is not very small

In this chapter we complete the proof of Theorem 1.2 with the following proposition.

PROPOSITION 19.1. *Suppose that Assumptions* 4.3 *and* 9.2 *hold, F is separating but not a semi-fibre, $t_1^+ = t_1^- = 0$, and $d > 1$. If $M(\alpha)$ is a small Seifert manifold but not very small, then $\Delta(\alpha, \beta) \leq 5$.*

We suppose that the hypotheses of the proposition hold for the remainder of this chapter. Since $d > 1$, M is not the figure eight knot exterior, and so $\Delta(\alpha, \beta) \leq 6$ by Proposition 15.1. We assume below that $\Delta(\alpha, \beta) = 6$ in order to derive a contradiction. Proposition 16.1 then implies that X^- is a twisted I-bundle.

19.1. Involutions on M and its fillings

For each $\epsilon \in \{+, -\}$, recall the essential annulus A^ϵ which separates $\widehat{X}^\epsilon$ into a union of two solid tori which contains $k_\epsilon = K_\beta \cap \widehat{X}^\epsilon$ as an essential arc. Recall as well that $\widehat{X}^+$ is Seifert fibred over $D^2(p_+, q_+)$ with $p_+ \geq 3$ and $q_+ \geq 2$, and $\widehat{X}^-$ is Seifert fibred over $D^2(2, 2)$.

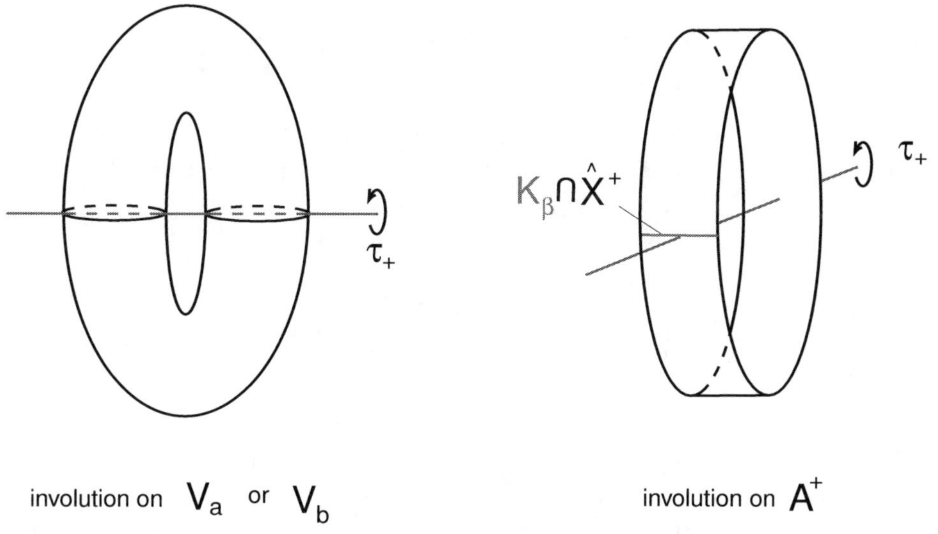

FIGURE 19.1. The involution τ_+

Let V_a and V_b be the two solid tori which arise as the components of $\widehat{X}^+ \setminus A^+$ where A^+ has winding number p_+ in V_a and q_+ in V_b.

There is an involution τ_+ on $\widehat{X}^+$ under which each of V_a, V_b, A^+, k_+ is invariant. More precisely,

- the restriction of τ_+ to both V_a and V_b is a standard involution of a solid torus whose fixed point set is a pair of arcs each contained in a meridional disk of the solid torus;
- the restriction of τ_+ to A^+ is a π-rotation with two fixed points (cf. Figure 19.1);
- the restriction of τ_+ to k_+ is a rotation about a fixed point (cf. Figure 19.1);
- the restriction of τ_+ to $\widehat{F}$ is a hyperelliptic involution which exchanges the two points ∂k_+;
- the quotient space of $\widehat{X}^+$ under τ_+ is a 3-ball whose branch set is shown in Figure 19.2(1).

There is an analogous involution τ_- on $\widehat{X}^-$ with quotient space as shown in Figure 19.2(2), at least up to homeomorphism.

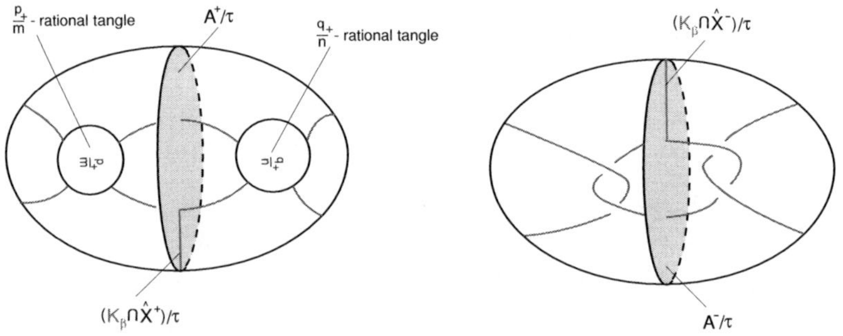

(1) (2)

FIGURE 19.2. The quotient spaces and branch sets

Let $f : \partial\widehat{X}^+ \to \partial\widehat{X}^-$ be the gluing map yielding $M(\beta) = \widehat{X}^+ \cup \widehat{X}^-$. Then as both $\tau_-|_{\widehat{F}}$ and $f \circ \tau_+|_{\widehat{F}} \circ f^{-1}$ are hyperelliptic involutions exchanging the two points $\partial k_+ = \partial k_-$, there is a homeomorphism g of $\widehat{X}^-$ isotopic (rel ∂k_-) to the identity such that

$$f \circ (\tau_+|_{\partial\widehat{X}^+}) \circ f^{-1} = (g \circ \tau_- \circ g^{-1})|_{\partial\widehat{X}^-}$$

Hence, up to replacing τ_- by $g \circ \tau_- \circ g^{-1}$, we may assume that the restriction of τ_+ to $(\partial\widehat{X}^+, k_+)$ equals that of τ_- to $(\partial\widehat{X}^-, k_-)$. Let

$$\tau_\beta : (M(\beta), K_\beta) \to (M(\beta), K_\beta)$$

be the involution obtained by gluing τ_+ and τ_- together along $\widehat{F}$.

We may assume that the filling solid torus V_β is τ_β-invariant and thus obtain an involution τ of $M = M(\beta) \setminus V_\beta$. The quotient spaces $M(\beta)/\tau_\beta$ and M/τ and their branch sets are as shown in Figure 19.3. The involution τ extends to an involution τ_γ on any γ-filling of M along ∂M and the branch set in $M(\gamma)/\tau_\gamma$ is obtained by filling M/τ with a corresponding rational tangle along $\partial M/\tau$. It follows that $M(\gamma)$ is a 2-fold cover of S^3 branch along a link L_γ.

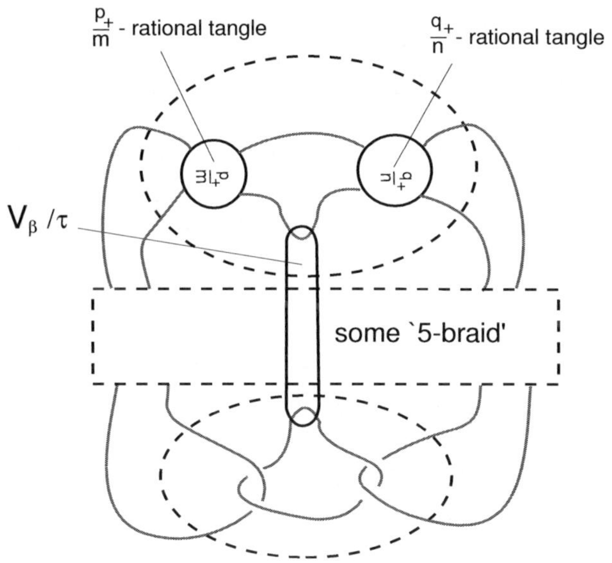

FIGURE 19.3. M/τ

19.2. Branch sets of the involutions on small Seifert manifolds

Suppose that W is Seifert fibred with base orbifold $S^2(a,b,c)$ where $a,b,c \geq 2$, but is not a prism manifold, which is the 2-fold cover of S^3 branched over a link L. By the orbifold theorem we may assume that the associated covering involution, θ say, preserves the Seifert structure on W. We distinguish two cases. If θ reverses the orientations of the Seifert fibres of W then L is a Montesinos link ([**Mo**]). If it preserves their orientations, the Seifert structure on W quotients to one on S^3 for which L is a union of fibres. In the former case we say that L is a Montesinos link L of *type* (a,b,c). In the latter case we say that L is a *Seifert link*.

Our goal in this section is to describe more precisely the possibilities for L when it is a Seifert link. We begin with some standard examples of Seifert links and the base orbifolds of their branched covers.

EXAMPLE 19.2. If L is a (p,q) torus knot, then its 2-fold branched cover has base orbifold

$$S^2(a,b,c) \cong \begin{cases} S^2(2,p,q) & \text{if } pq \text{ is odd} \\ S^2(\frac{p}{2},q,q) & \text{if } p \text{ is even} \\ S^2(\frac{q}{2},p,p) & \text{if } q \text{ is even} \end{cases}$$

EXAMPLE 19.3. If $L = U \cup K \subset S^3$ is a two component link where U is an unknot and K is a (p,q)-cable of U (i.e. K is isotopic in $S^3 - U$ into the boundary of a regular neighbourhood $N(U)$ of U as a (p,q)-torus knot with winding number q in $N(U)$), then we say K is a *(p,q)-torus knot with respect to U*. In this case, the linking number between U and K is p. The 2-fold branched cover of L has base orbifold

$$S^2(a,b,c) \cong \begin{cases} S^2(q,p,p) & \text{if } q \text{ is odd} \\ S^2(2,p,2q) & \text{if } q \text{ is even} \end{cases}$$

EXAMPLE 19.4. If $L = U_0 \cup U_1 \cup K \subset S^3$ is a three component link where $U_0 \cup U_1$ is a Hopf link and K is a (p,q)-torus knot which is isotopic in $S^3 - (U_0 \cup U_1)$ into the boundary of a regular neighbourhood $N(U_0)$ of U_0, then we say that K is a *(p,q)-torus knot with respect to* $U_0 \cup U_1$. The 2-fold branched cover of L has base orbifold

$$S^2(a,b,c) \cong S^2(2,2p,2q)$$

The following lemma is a consequence of [**BGZ3**, Lemma 4.3] and the list of the base orbifolds of the 2-fold branched covers of the Seifert links in the three examples above.

LEMMA 19.5. *Let W and θ be as above. Suppose that θ preserves the orientations of the Seifert fibres of W and endow W/τ with the induced Seifert structure. Assume as well that $W/\theta = S^3$ and let $L_\theta \subset W/\tau \cong S^3$ be the branch set of θ.*

(1) *L_θ is a union of at most three Seifert fibres of S^3. Moreover at least one component of L_θ is a regular fibre.*
(2) *If $|L_\theta| = 1$, then either*
   (i) *$(a,b,c) = (2,p,q)$ where p and q are odd, $\gcd(p,q) = 1$, and L_θ is a (p,q) torus knot, or*
   (ii) *$(a,b,c) = (p/2,q,q)$ where p is even, $\gcd(p,q) = 1$, and L_θ is a (p,q)-torus knot.*
(3) *If $|L_\theta| = 2$, then either*
   (i) *$(a,b,c) = (2,p,2q)$ where q is even, $\gcd(p,q) = 1$, and L_θ is the union of an unknot U*
   and a (p,q)-torus knot with respect to U, or
   (ii) *$(a,b,c) = (q,p,p)$ where q is odd, $\gcd(p,q) = 1$, and L_θ is the union of an unknot U*
   and a (p,q)-torus knot with respect to U.
(4) *If $|L_\theta| = 3$, then $(a,b,c) = (2,2p,2q)$ where $\gcd(p,q) = 1$ and L_θ is the union of a Hopf link and a (p,q)-torus knot K with respect to $L_\theta - K$.* ◇

19.3. The proof of Proposition 19.1

Suppose that $M(\alpha)$ is Seifert fibred with base orbifold $S^2(a,b,c)$ where $a,b,c \geq 2$, but is not a prism manifold. We know that with respect to the involution τ_α constructed in §19.1, $M(\alpha)$ is a 2-fold branched cover of (S^3, L_α) where L_α is either a Montesinos link or a link whose exterior is Seifert fibred.

As $\Delta(\alpha,\beta) = 6$ is even, L_α contains a trivial component U_0, as shown in Figure 19.4(1), where $L_\alpha - U_0$ is a Montesinos link whose 2-fold branched cover is obtained by Dehn filling $\widehat{X}^+$ with a slope of distance d from ϕ_+. Thus $L_\alpha - U_0$ is a Montesinos link of type (p_+, q_+, d) (cf. Figure 19.4(2)). As $d > 1$, the fundamental group of the 2-fold branched cover of $L_\alpha - U_0$ is not cyclic, so $L_\alpha - U_0$ is not a 2-bridge link, and therefore L_α cannot be a Montesinos link of three cyclically composed rational tangles. Hence L_α must be a Seifert link of two or three components. In particular, τ_α can be assumed to preserve the orientations of the Seifert fibres of

$M(\alpha)$, so that Lemma 19.5 applies to our situation. Then $L_\alpha - U_0$ has either one or two components.

CASE 1. $K = L_\alpha - U_0$ is a knot. If $K = L_\alpha - U_0$ is a knot, it is both a torus knot and a Montesinos knot of type (p_+, q_+, d). Therefore K is either the $(-2,3,3)$-pretzel knot or the $(-2,3,5)$-pretzel knot ([**Oe**]). Equivalently, K is either the $(3,4)$-torus knot or, respectively, the $(3,5)$-torus knot. Further, the linking number between U_0 and K is an odd number (see Figure 19.4(2) and apply [**BuZi**, Remark 12.7]).

First suppose that $K = T(3,4)$. Then K is a $(3,4)$-torus knot with respect to U_0. Lemma 19.5 then shows that $M(\alpha)$ has base orbifold $S^2(2,8,3)$. On the other hand, the 2-fold branched cover of K has base orbifold $S^2(p_+, q_+, d) \cong S^2(2,3,3)$, so (p_+, q_+, d) is either $(2,3,3)$ or $(3,3,2)$.

If $(p_+, q_+, d) = (2,3,3)$, there is an epimorphism $\pi_1(M(\beta)) \xrightarrow{\varphi} \Delta(2,3,3) *_\psi \Delta(2,2,3)$ and as in Lemma 18.2, we can build a strictly non-trivial curve X_0 in $X_{PSL_2}(M(\beta)) \subset X_{PSL_2}(M)$ which contains only irreducible characters and for which $s_{X_0} \geq 2$. Further, $\tilde{f}_\alpha$ has a pole at each of ideal point of X_0. By [**BeBo**, Proposition 3.2], $\Delta(2,8,3)$ has exactly four irreducible characters and therefore Proposition 5.2(2) implies that $\Delta(\alpha,\beta) \leq 1 + \frac{8}{2} = 5$, contrary to our assumptions.

If $(p_+, q_+, d) = (3,3,2)$, there is an epimorphism $\pi_1(M(\beta)) \xrightarrow{\varphi} \Delta(3,3,2) *_\psi \Delta(2,2,2)$ and as above we can build a strictly non-trivial curve X_0 in $X_{PSL_2}(M(\beta)) \subset X_{PSL_2}(M)$ which contains only irreducible characters and for which $s_{X_0} \geq 1$. Further, $\tilde{f}_\alpha$ has a pole at each of ideal point of X_0. Two of the four irreducible characters of $\Delta(2,8,3)$ correspond to the discrete faithful representation $\Delta(2,8,3) \to PSL_2(\mathbb{R}) \to PSL_2(\mathbb{C})$ and to the quotient map to $\Delta(2,2,3)$. Neither of these can lie on X_0, so Proposition 5.2(2) implies that $\Delta(\alpha,\beta) \leq 1 + \frac{4}{1} = 5$, contrary to our assumptions. Thus $K \neq T(3,4)$.

Next suppose that $K = T(3,5)$. By Lemma 19.5, the base orbifold of $M(\alpha)$ is either $S^2(3,3,5)$ or $S^2(3,5,5)$. Further, the base orbifold of the 2-fold branched cover of K is $S^2(2,3,5)$, so $(p_+, q_+, d) = (2,3,5)$ or $(2,5,3)$ or $(3,5,2)$. In any event, we can construct a strictly non-trivial curve $X_0 \subset X_{PSL_2}(M)$ as above. In fact, since there are two conjugacy classes of elements of order 5 in $PSL_2(\mathbb{C})$, we can construct disjoint, strictly non-trivial curve $X_0, X_1 \subset X_{PSL_2}(M)$. Since $\Delta(3,3,5)$ and $\Delta(3,5,5)$ are generated by elements of odd order, no representation whose character lies on these curves can have image in $\mathcal{N}$. Hence if $s_{X_j} = 1$ for some j, Proposition 5.1(3) implies that $\Delta(\alpha,\beta)$ is odd, contrary to assumption. Thus $s_{X_j} \geq 2$ for both j. Set $X = X_0 \cup X_1$. Then $\|\cdot\|_X = \|\cdot\|_{X_0} + \|\cdot\|_{X_1}$ so that $s_X = s_{X_0} + s_{X_1} \geq 4$. Since $\Delta(3,3,5)$ has exactly four irreducible characters and $\Delta(3,5,5)$ exactly eight ([**BeBo**, Proposition 3.2]), we see that $\Delta(\alpha,\beta) \leq 1 + \frac{16}{4} = 5$, a contradiction. We have therefore ruled out the possibility that $L_\alpha - U_0$ is a knot.

CASE 2. $L_\alpha - U_0$ is a two component link. If $L_\alpha - U_0$ is a two component link, then $L_\alpha - U_0 = U_1 \cup K_1$ where U_1 is a trivial knot and K_1 is both a 2-bridge knot and a torus knot. Thus K_1 is a $(2,n)$ torus knot for some odd integer n. By Lemma 19.5(3), $M(\alpha)$ is Seifert fibred over $S^2(2,4,2q)$.

Now K is either a $(2,n)$ or $(n,2)$ torus knot with respect to U_1. Hence the base orbifold $S^2(p_+, q_+, d)$ of the 2-fold branched cover of $U_1 \cup K_1$ is either $S^2(2,4,n)$ or $S^2(2,2,n)$, so exactly two of p_+, q_+ and d are even. Now d cannot be odd, as otherwise p_+ and q_+ would both be even. Inspection of Figure 19.4(3) then shows

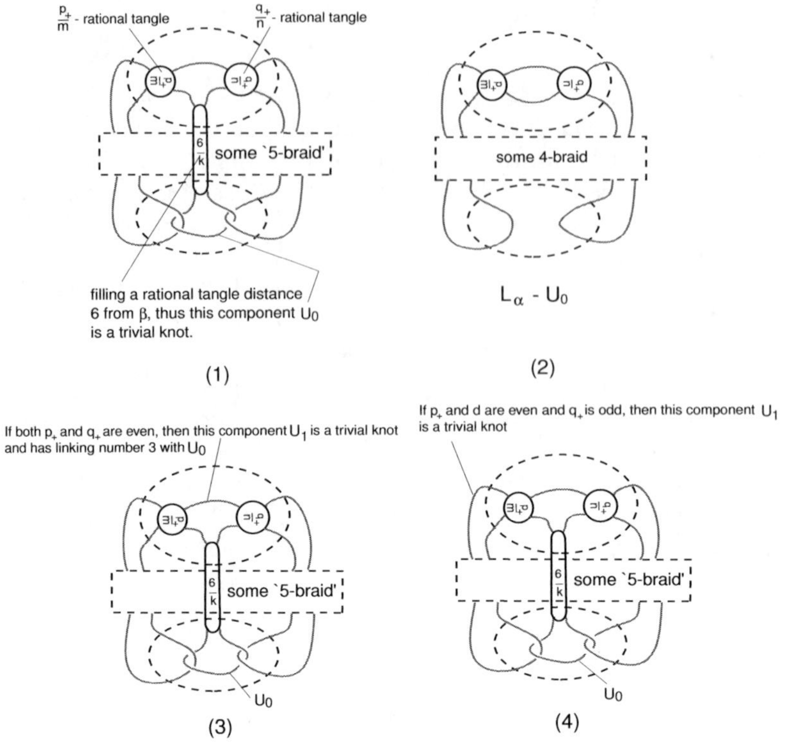

FIGURE 19.4. Possible branch sets for $6/k$-tangle fillings

that the linking number between U_0 and U_1 is 3, which is impossible since $U_1 \cup U_0$ is a Hopf link. We assume then that p_+ and d are even. The case that q_+ and d are even is treated similarly.

Since $q_+ = n$ odd, we know that (p_+, q_+, d) is either $(2, q_+, 2), (2, q_+, 4)$ or $(4, q_+, 2)$. Further, U_1 is as illustrated in Figure 19.4(4) and K_1 is a $(2, q_+)$-torus knot. We can also see from the figure that the linking number between U_0 and K_1 is even and therefore, K_1 is a $(q_+, 2)$-torus knot with respect to U_1. Hence the base orbifold $S^2(p_+, q_+, d)$ of the 2-fold branched cover of $U_1 \cup K_1$ is $S^2(2, 4, q_+)$, so $(p_+, q_+, d) = (2, q_+, 4)$ or $(4, q_+, 2)$.

If $d = 4$, we can build $\lfloor \frac{q_+}{2} \rfloor$ strictly non-trivial disjoint curves X_0 as above, each with $s_{X_0} \geq 2$ (cf. §6.1). Hence if X is the union of these curves, $s_X \geq q_+ - 1$. Similarly if $d = 2$, we can build $2\lfloor \frac{q_+}{2} \rfloor = q_+ - 1$ strictly non-trivial disjoint curves X_0 as above, each with $s_{X_0} \geq 1$. Hence if X is the union of these curves, we again have $s_X \geq q_+ - 1$. By Lemma 19.5(2), $M(\alpha)$ is Seifert fibred over $S^2(2, 4, 2q_+)$ and by [**BeBo**, Proposition 3.2]), $\Delta(2, 4, 2q_+)$ has exactly q_+ irreducible characters of representations with values in $PSL_2(\mathbb{C})$. Hence, $\Delta(\alpha, \beta) \leq 1 + \frac{2q_+}{q_+ - 1} \leq 5$, contrary to our assumptions. This rules out the possibility that $L_\alpha - U_0$ is a two component link and completes the proof of Proposition 19.1.

CHAPTER 20

Proof of Theorem 1.8

The first proposition deals with the case where $M(\alpha)$ is of C- or D-type.

PROPOSITION 20.1. *Let M be a hyperbolic knot manifold with slopes α and β on ∂M such that $M(\beta)$ is toroidal and $M(\alpha)$ is a very small Seifert manifold.*

(1) *If $M(\alpha)$ is of C-type then $\Delta(\alpha, \beta) \leq 4$.*
(2) *If $M(\alpha)$ is of D-type then*
    (a) *$\Delta(\alpha, \beta) \leq 4$, and*
    (b) *if $M(\beta)$ is a torus semi-bundle then $\Delta(\alpha, \beta) \leq 3$.*

PROOF OF 20.1.
(1) If $M(\alpha)$ is S^3 then $\Delta(\alpha, \beta) \leq 2$ by [**GL1**], if $M(\alpha)$ is $S^1 \times S^2$ then $\Delta(\alpha, \beta) \leq 3$ by [**Oh**] and [**Wu2**], and if $M(\alpha)$ is a lens space then $\Delta(\alpha, \beta) \leq 4$ by [**L3**].

(2)(a) If M is a hyperbolic knot manifold such that $M(\beta)$ is toroidal and $M(\alpha)$ contains a Klein bottle, it follows from [**V**] (if $\Delta(\alpha, \beta) > 5$) and [**L2**] (if $\Delta(\alpha, \beta) = 5$), that $M(\alpha)$ is toroidal. If $M(\alpha)$ is of D-type then it contains a Klein bottle but is atoroidal. Hence $\Delta(\alpha, \beta) \leq 4$. Part (b) is Proposition 7.2(5). ◊

PROPOSITION 20.2. *Let M be a hyperbolic knot manifold which contains an m-punctured torus F with boundary slope β, that is not a fibre in M if $m \geq 3$. Let α be a slope on ∂M such that $M(\alpha)$ is a very small Seifert manifold. Then*

(1) *$\Delta(\alpha, \beta) \leq 5$, and*
(2) *if $\Delta(\alpha, \beta) = 5$, then either*
    (a) *$m = 1$, F is not a fibre, $(M; \alpha, \beta) \cong (Wh(-3/2); -5, 0)$, and $M(\alpha)$ has base orbifold $S^2(2, 3, 3)$, or*
    (b) *F is separating in M though not a semi-fibre.*

PROOF. This follows from [**BGZ2**, Propositions 13.1 and 13.2 and Theorem 2.7], [**BGZ3**, Theorem 1.3], Theorem 1.2, Proposition 10.1, and the fact that the figure eight knot exterior admits no small Seifert Dehn fillings which are very small. ◊

Theorem 1.8 follows from Propositions 20.1, 7.2, and 20.2.

Bibliography

[Ad] Colin C. Adams, *The noncompact hyperbolic 3-manifold of minimal volume*, Proc. Amer. Math. Soc. **100** (1987), no. 4, 601–606, DOI 10.2307/2046691. MR894423

[BeBo] Leila Ben Abdelghani and Steven Boyer, *A calculation of the Culler-Shalen seminorms associated to small Seifert Dehn fillings*, Proc. London Math. Soc. (3) **83** (2001), no. 1, 235–256, DOI 10.1112/plms/83.1.235. MR1829566

[BoiBo] Michel Boileau and Steven Boyer, *On character varieties, sets of discrete characters, and nonzero degree maps*, Amer. J. Math. **134** (2012), no. 2, 285–347, DOI 10.1353/ajm.2012.0013. MR2904999

[BoiZie] Michel Boileau and Heiner Zieschang, *Heegaard genus of closed orientable Seifert 3-manifolds*, Invent. Math. **76** (1984), no. 3, 455–468, DOI 10.1007/BF01388469. MR746538

[Bo] Steven Boyer, *On the local structure of* $\mathrm{SL}(2,\mathbb{C})$*-character varieties at reducible characters*, Topology Appl. **121** (2002), no. 3, 383–413, DOI 10.1016/S0166-8641(01)00278-4. MR1909000

[BCSZ1] Steve Boyer, Marc Culler, Peter B. Shalen, and Xingru Zhang, *Characteristic subsurfaces and Dehn filling*, Trans. Amer. Math. Soc. **357** (2005), no. 6, 2389–2444, DOI 10.1090/S0002-9947-04-03576-7. MR2140444

[BCSZ2] Steve Boyer, Marc Culler, Peter B. Shalen, and Xingru Zhang, *Characteristic subsurfaces, character varieties and Dehn fillings*, Geom. Topol. **12** (2008), no. 1, 233–297, DOI 10.2140/gt.2008.12.233. MR2390346

[BGZ1] Steven Boyer, Cameron McA. Gordon, and Xingru Zhang, *Dehn fillings of large hyperbolic 3-manifolds*, J. Differential Geom. **58** (2001), no. 2, 263–308. MR1913944

[BGZ2] Steven Boyer, Cameron McA. Gordon, and Xingru Zhang, *Characteristic submanifold theory and toroidal Dehn filling*, Adv. Math. **230** (2012), no. 4-6, 1673–1737, DOI 10.1016/j.aim.2012.03.029. MR2927352

[BGZ3] Steven Boyer, Cameron McA. Gordon, and Xingru Zhang, *Dehn fillings of knot manifolds containing essential once-punctured tori*, Trans. Amer. Math. Soc. **366** (2014), no. 1, 341–393, DOI 10.1090/S0002-9947-2013-05837-0. MR3118399

[BZ1] Steven Boyer and Xingru Zhang, *Finite Dehn surgery on knots*, J. Amer. Math. Soc. **9** (1996), no. 4, 1005–1050, DOI 10.1090/S0894-0347-96-00201-9. MR1333293

[BZ2] Steven Boyer and Xingru Zhang, *On Culler-Shalen seminorms and Dehn filling*, Ann. of Math. (2) **148** (1998), no. 3, 737–801, DOI 10.2307/121031. MR1670053

[BuZi] Gerhard Burde and Heiner Zieschang, *Knots*, De Gruyter Studies in Mathematics, vol. 5, Walter de Gruyter & Co., Berlin, 1985. MR808776

[CDGW] Marc Culler, Nathan M. Dunfield, Matthias Goemer, and Jeffrey R. Weeks, *SnapPy, a computer program for studying the geometry and topology of 3-manifolds*, `https://snappy.computop.org`, (date accessed).

[CGLS] Marc Culler, Cameron McA. Gordon, John Luecke, and Peter B. Shalen, *Dehn surgery on knots*, Ann. of Math. (2) **125** (1987), no. 2, 237–300, DOI 10.2307/1971311. MR881270

[Ga] David Gabai, *Foliations and the topology of 3-manifolds. II*, J. Differential Geom. **26** (1987), no. 3, 461–478. MR910017

[Go1] Cameron McA. Gordon, *Boundary slopes of punctured tori in 3-manifolds*, Trans. Amer. Math. Soc. **350** (1998), no. 5, 1713–1790, DOI 10.1090/S0002-9947-98-01763-2. MR1390037

[Gor98] C. McA. Gordon, *Dehn filling: a survey*, Knot theory (Warsaw, 1995), Banach Center Publ., vol. 42, Polish Acad. Sci. Inst. Math., Warsaw, 1998, pp. 129–144. MR1634453

[GL1] Cameron McA. Gordon and John Luecke, *Dehn surgeries on knots creating essential tori. I*, Comm. Anal. Geom. **3** (1995), no. 3-4, 597–644, DOI 10.4310/CAG.1995.v3.n4.a3. MR1371211

[GL2] Cameron McA. Gordon and John Luecke, *Non-integral toroidal Dehn surgeries*, Comm. Anal. Geom. **12** (2004), no. 1-2, 417–485. MR2074884

[GW] Cameron McA. Gordon and Ying-Qing Wu, *Toroidal Dehn fillings on hyperbolic 3-manifolds*, Mem. Amer. Math. Soc. **194** (2008), no. 909, vi+140, DOI 10.1090/memo/0909. MR2419168

[GW2] Cameron McA. Gordon and Ying-Qing Wu, *Toroidal and annular Dehn fillings*, Proc. London Math. Soc. (3) **78** (1999), no. 3, 662–700, DOI 10.1112/S0024611599001823. MR1674841

[Ha] Allen Hatcher, *Notes on basic* 3*-manifold topology*, Course notes, available at `https://pi.math.cornell.edu/~hatcher`.

[He] John Hempel, 3*-Manifolds*, Princeton University Press, Princeton, N. J.; University of Tokyo Press, Tokyo, 1976. Ann. of Math. Studies, No. 86. MR0415619

[Ja] William Jaco, *Lectures on three-manifold topology*, CBMS Regional Conference Series in Mathematics, vol. 43, American Mathematical Society, Providence, R.I., 1980. MR565450

[Ja2] William Jaco, *Adding a 2-handle to a 3-manifold: an application to property R*, Proc. Amer. Math. Soc. **92** (1984), no. 2, 288–292, DOI 10.2307/2045205. MR754723

[Ko] Tsuyoshi Kobayashi, *Structures of the Haken manifolds with Heegaard splittings of genus two*, Osaka J. Math. **21** (1984), no. 2, 437–455. MR752472

[L1] Sangyop Lee, *Dehn fillings yielding Klein bottles*, Int. Math. Res. Not., posted on 2006, Art. ID 24253, 34, DOI 10.1155/IMRN/2006/24253. MR2219228

[L2] Sangyop Lee, *Klein bottle and toroidal Dehn fillings at distance 5*, Pacific J. Math. **247** (2010), no. 2, 407–434, DOI 10.2140/pjm.2010.247.407. MR2734156

[L3] Sangyop Lee, *Lens spaces and toroidal Dehn fillings*, Math. Z. **267** (2011), no. 3-4, 781–802, DOI 10.1007/s00209-009-0646-0. MR2776058

[LM] Alexander Lubotzky and Andy R. Magid, *Varieties of representations of finitely generated groups*, Mem. Amer. Math. Soc. **58** (1985), no. 336, xi+117, DOI 10.1090/memo/0336. MR818915

[Ma] Bruno Martelli, *A Python program for computing exceptional fillings on a finite volume cusped hyperbolic* 3*-manifold*, available at his web site: `people.dm.unipi.it/martelli/Dehn.html`

[MP] Bruno Martelli and Carlo Petronio, *Dehn filling of the "magic" 3-manifold*, Comm. Anal. Geom. **14** (2006), no. 5, 969–1026. MR2287152

[MS] Daniel Matignon and Nabil Sayari, *Klein slopes on hyperbolic 3-manifolds*, Math. Proc. Cambridge Philos. Soc. **143** (2007), no. 2, 419–447, DOI 10.1017/S0305004107000229. MR2364660

[Mo] José M. Montesinos, *Seifert manifolds that are ramified two-sheeted cyclic coverings* (Spanish), Bol. Soc. Mat. Mexicana (2) **18** (1973), 1–32. MR341467

[Oe] Ulrich Oertel, *Closed incompressible surfaces in complements of star links*, Pacific J. Math. **111** (1984), no. 1, 209–230. MR732067

[Oh] Seungsang Oh, *Reducible and toroidal* 3*-manifolds obtained by Dehn fillings*, Topology Appl. **75** (1997), no. 1, 93–104, DOI 10.1016/S0166-8641(96)00065-X. MR1425387

[Rlf] Dale Rolfsen, *Knots and links*, Mathematics Lecture Series, No. 7, Publish or Perish, Inc., Berkeley, Calif., 1976. MR0515288

[Rou] Fionntan Wouter Murray Roukema, *Exceptional slopes on manifolds of small complexity*, Comm. Anal. Geom. **27** (2019), no. 5, 1105–1161, DOI 10.4310/CAG.2019.v27.n5.a4. MR4037377

[S] Eric Sedgwick, *The irreducibility of Heegaard splittings of Seifert fibered spaces*, Pacific J. Math. **190** (1999), no. 1, 173–199, DOI 10.2140/pjm.1999.190.173. MR1722770

[V] Luis G. Valdez-Sánchez, *Toroidal and Klein bottle boundary slopes*, Topology Appl. **154** (2007), no. 3, 584–603, DOI 10.1016/j.topol.2006.08.001. MR2280903

[Wu1] Ying Qing Wu, *Incompressibility of surfaces in surgered* 3*-manifolds*, Topology **31** (1992), no. 2, 271–279, DOI 10.1016/0040-9383(92)90020-I. MR1167169

[Wu2] Ying-Qing Wu, *Dehn fillings producing reducible manifolds and toroidal manifolds*, Topology **37** (1998), no. 1, 95–108, DOI 10.1016/S0040-9383(97)00016-5. MR1480879

[Wu3] Ying-Qing Wu, *The classification of Dehn fillings on the outer torus of a 1-bridge braid exterior which produce solid tori*, Math. Ann. **330** (2004), no. 1, 1–15, DOI 10.1007/s00208-004-0519-0. MR2091675

Editorial Information

To be published in the *Memoirs*, a paper must be correct, new, nontrivial, and significant. Further, it must be well written and of interest to a substantial number of mathematicians. Piecemeal results, such as an inconclusive step toward an unproved major theorem or a minor variation on a known result, are in general not acceptable for publication.

Papers appearing in *Memoirs* are generally at least 80 and not more than 200 published pages in length. Papers less than 80 or more than 200 published pages require the approval of the Managing Editor of the Transactions/Memoirs Editorial Board. Published pages are the same size as those generated in the style files provided for AMS-LaTeX.

Information on the backlog for this journal can be found on the AMS website starting from `http://www.ams.org/memo`.

A Consent to Publish is required before we can begin processing your paper. After a paper is accepted for publication, the Providence office will send a Consent to Publish and Copyright Agreement to all authors of the paper. By submitting a paper to the *Memoirs*, authors certify that the results have not been submitted to nor are they under consideration for publication by another journal, conference proceedings, or similar publication.

Information for Authors

Memoirs is an author-prepared publication. Once formatted for print and on-line publication, articles will be published as is with the addition of AMS-prepared frontmatter and backmatter. Articles are not copyedited; however, confirmation copy will be sent to the authors.

Initial submission. The AMS uses Centralized Manuscript Processing for initial submissions. Authors should submit a PDF file using the Initial Manuscript Submission form found at `www.ams.org/submission/memo`, or send one copy of the manuscript to the following address: Centralized Manuscript Processing, MEMOIRS OF THE AMS, 201 Charles Street, Providence, RI 02904-2294 USA. If a paper copy is being forwarded to the AMS, indicate that it is for *Memoirs* and include the name of the corresponding author, contact information such as email address or mailing address, and the name of an appropriate Editor to review the paper (see the list of Editors below).

The paper must contain a *descriptive title* and an *abstract* that summarizes the article in language suitable for workers in the general field (algebra, analysis, etc.). The *descriptive title* should be short, but informative; useless or vague phrases such as "some remarks about" or "concerning" should be avoided. The *abstract* should be at least one complete sentence, and at most 300 words. Included with the footnotes to the paper should be the 2020 *Mathematics Subject Classification* representing the primary and secondary subjects of the article. The classifications are accessible from `www.ams.org/msc/`. The Mathematics Subject Classification footnote may be followed by a list of *key words and phrases* describing the subject matter of the article and taken from it. Journal abbreviations used in bibliographies are listed in the latest *Mathematical Reviews* annual index. The series abbreviations are also accessible from `www.ams.org/msnhtml/serials.pdf`. To help in preparing and verifying references, the AMS offers MR Lookup, a Reference Tool for Linking, at `www.ams.org/mrlookup/`.

Electronically prepared manuscripts. The AMS encourages electronically prepared manuscripts, with a strong preference for AMS-LaTeX. To this end, the Society has prepared AMS-LaTeX author packages for each AMS publication. Author packages include instructions for preparing electronic manuscripts, samples, and a style file that generates the particular design specifications of that publication series.

Authors may retrieve an author package for *Memoirs of the AMS* from `www.ams.org/journals/memo/memoauthorpac.html`. The *AMS Author Handbook* is available in PDF format from the author package link. The author package can also be obtained free of charge by sending email to `tech-support@ams.org` or from the Publication Division,

American Mathematical Society, 201 Charles St., Providence, RI 02904-2294, USA. When requesting an author package, please specify the publication in which your paper will appear. Please be sure to include your complete mailing address.

After acceptance. The source files for the final version of the electronic manuscript should be sent to the Providence office immediately after the paper has been accepted for publication. The author should also submit a PDF of the final version of the paper to the editor, who will forward a copy to the Providence office.

Accepted electronically prepared files can be submitted via the web at `www.ams.org/submit-book-journal/`, sent via FTP, or sent on CD to the Electronic Prepress Department, American Mathematical Society, 201 Charles Street, Providence, RI 02904-2294 USA. TEX source files and graphic files can be transferred over the Internet by FTP to the Internet node `ftp.ams.org` (130.44.1.100). When sending a manuscript electronically via CD, please be sure to include a message indicating that the paper is for the *Memoirs*.

Electronic graphics. Comprehensive instructions on preparing graphics are available at `www.ams.org/authors/journals.html`. A few of the major requirements are given here.

Submit files for graphics as EPS (Encapsulated PostScript) files. This includes graphics originated via a graphics application as well as scanned photographs or other computer-generated images. If this is not possible, TIFF files are acceptable as long as they can be opened in Adobe Photoshop or Illustrator.

Authors using graphics packages for the creation of electronic art should also avoid the use of any lines thinner than 0.5 points in width. Many graphics packages allow the user to specify a "hairline" for a very thin line. Hairlines often look acceptable when proofed on a typical laser printer. However, when produced on a high-resolution laser imagesetter, hairlines become nearly invisible and will be lost entirely in the final printing process.

Screens should be set to values between 15% and 85%. Screens which fall outside of this range are too light or too dark to print correctly. Variations of screens within a graphic should be no less than 10%.

Any graphics created in color will be rendered in grayscale for the printed version unless color printing is authorized by the Managing Editor and the Publisher. In general, color graphics will appear in color in the online version.

Inquiries. Any inquiries concerning a paper that has been accepted for publication should be sent to `memo-query@ams.org` or directly to the Electronic Prepress Department, American Mathematical Society, 201 Charles St., Providence, RI 02904-2294 USA.

Editors

SELECTED PUBLISHED TITLES IN THIS SERIES

1468 **Alice Hedenlund and John Rognes,** A Multiplicative Tate Spectral Sequence for Compact Lie Group Actions, 2024

1467 **Qi Lü and Xu Zhang,** Optimal Feedback for Stochastic Linear Quadratic Control and Backward Stochastic Riccati Equations in Infinite Dimensions, 2024

1466 **Martin W. Liebeck, Gary M. Seitz, and Donna M. Testerman,** Multiplicity-free Representations of Algebraic Groups, 2024

1465 **Tobias Rossmann and Christopher Voll,** Groups, Graphs, and Hypergraphs: Average Sizes of Kernels of Generic Matrices with Support Constraints, 2024

1464 **Jens Marklof and Andreas Strömbergsson,** Kinetic Theory for the Low-Density Lorentz Gas, 2024

1463 **You Qi and Joshua Sussan,** p-DG Cyclotomic nilHecke Algebras II, 2024

1462 **Mikhail Khovanov, You Qi, and Joshua Sussan,** p-DG Cyclotomic nilHecke Algebras, 2024

1461 **James Gabe,** Classification of $\mathcal{O}_\infty$-Stable C^*-Algebras, 2024

1460 **Erik Bates,** Empirical Measures, Geodesic Lengths, and a Variational Formula in First-Passage Percolation, 2024

1459 **Jonathan Brundan and Catharina Stroppel,** Semi-Infinite Highest Weight Categories, 2024

1458 **Siddhant Agrawal,** Angled Crested Like Water Waves with Surface Tension II: Zero Surface Tension Limit, 2024

1457 **Paul-Elliot Anglès d'Auriac, Peter A. Cholak, Damir D. Dzhafarov, Benoît Monin, and Ludovic Patey,** Milliken's Tree Theorem and Its Applications: A Computability-Theoretic Perspective, 2024

1456 **Paul Gartside and Max Pitz,** Eulerian Spaces, 2023

1455 **Francisco Gancedo, Eduardo García-Juárez, Neel Patel, and Robert M. Strain,** Global Regularity for Gravity Unstable Muskat Bubbles, 2023

1454 **Michael Handel and Lee Mosher,** Hyperbolic Actions and 2nd Bounded Cohomology of Subgroups of $\mathrm{Out}(F_n)$, 2023

1453 **Yang Li,** SYZ Geometry for Calabi-Yau 3-folds: Taub-NUT and Ooguri-Vafa Type Metrics, 2023

1452 **Chris Parker, Gerald Pientka, Andreas Seidel, and Gernot Stroth,** Finite Groups Which Are Almost Groups of Lie Type in Characteristic **p**, 2023

1451 **Gili Golan Polak,** The Generation Problem in Thompson Group F, 2023

1450 **Mathias Pétréolle, Alan D. Sokal, and Bao-Xuan Zhu,** Lattice Paths and Branched Continued Fractions: An Infinite Sequence of Generalizations of the Stieltjes–Rogers and Thron–Rogers Polynomials, with Coefficientwise Hankel-Total Positivity, 2023

1449 **Quoc-Hung Nguyen,** Potential Estimates and Quasilinear Parabolic Equations with Measure Data, 2023

1448 **S. Launois, T. H. Lenagan, and B. M. Nolan,** Total Positivity is a Quantum Phenomenon: The Grassmannian Case, 2023

1447 **Andrew Lawrie, Jonas Lührmann, Sung-Jin Oh, and Sohrab Shahshahani,** Local Smoothing Estimates for Schrödinger Equations on Hyperbolic Space, 2023

1446 **Camillo De Lellis, Guido De Philippis, Jonas Hirsch, and Annalisa Massaccesi,** On the Boundary Behavior of Mass-Minimizing Integral Currents, 2023

1445 **K. R. Goodearl and M. T. Yakimov,** Cluster Algebra Structures on Poisson Nilpotent Algebras, 2023

1444 **Zev Rosengarten,** Tate Duality in Positive Dimension over Function Fields, 2023

For a complete list of titles in this series, visit the
AMS Bookstore at **www.ams.org/bookstore/memoseries/**.